SpringerBriefs on Pioneers in Science and Practice

Volume 1

Series Editor

Hans Günter Brauch

T0401989

For further volumes:
http://www.springer.com/series/10970
http://www.afes-press-books.de/html/SpringerBriefs_PSP.htm

Arthur H. Westing

Arthur H. Westing

Pioneer on the Environmental Impact of War

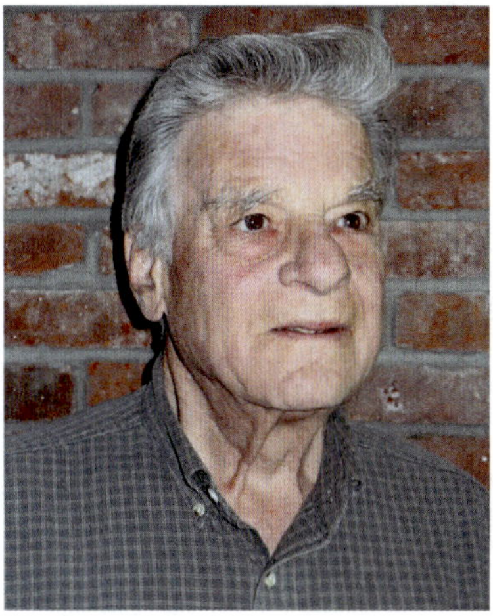

Springer

Arthur H. Westing
Westing Associates in Environment
Security, and Education
Putney, VT
USA

ISSN 2194-3125 ISSN 2194-3133 (electronic)
ISBN 978-3-642-31321-9 ISBN 978-3-642-31322-6 (eBook)
DOI 10.1007/978-3-642-31322-6
Springer Heidelberg New York Dordrecht London

Library of Congress Control Number: 2012941417

Printed on acid-free paper

Springer is part of Springer Science+Business Media (www.springer.com)

Editorial

SpringerBriefs on Pioneers in Science and Practice (PSP)[1]

Hans Günter Brauch[2]

This series presents eminent conceptual thinkers, scholars, policymakers, and diplomats who—as pioneers in a specific field of research or an area of political debate—had an innovative, longlasting regional or global impact on issues crucial for humankind in the twenty-first century.

This series will present the work of distinguished scientists in the natural or social sciences and in the humanities who inspired scientific and policy debates—from many disciplines and from all parts of the world and who with their conceptual and scholarly work have introduced new areas or topics of scientific analysis, research and teaching. This series on Pioneers in Science and Practice will also include influential and successful policymakers who have had a major impact on multilateral diplomacy and decision-making primarily in the four key areas of the environment, security, development, and peace.

The goal of this series on pioneers is to publish selected major texts by a pioneer in a particular field of scientific analysis and political practice together with a biography and a bibliography—or for practitioners a survey of major policy decisions they have influenced. The reader who is interested in the impact of new ideas on scientific development in different disciplines and in innovative policy decisions on international policies can thus easily access the overall work of the pioneer presented.

---

1 For a list of all titles in this series see at: http://www.afes-press-books.de/html/SpringerBriefs_PSP.htm.

2 Hans Günter Brauch (Dr. Phil., Heidelberg University; Dr. habil., Free University of Berlin, Germany) has taught as a Privatdozent (Adj. Prof.) at the Otto-Suhr Institute for Political Science of the Free University of Berlin since 1999; since 1987 he has been Chairman of Peace Research European Security Studies (AFES-PRESS), an independent non-governmental and tax-exempt international scientific society in Mosbach, Germany. Since 2003 he has been editor of the Hexagon Series on Human and Environmental Security and Peace (HESP) and since 2012 of the new SpringerBriefs in Environment, Security, Development and Peace (ESDP) and of these SpringerBriefs in Pioneers in Science and Practice (PSP), all published by Springer. He also teaches at SciencePo in Paris, at the European Peace University (EPU) in Schlaining (Austria), and on the Ph.D. Programme of the Centro de Estudios Superiores Navales (CESNAV) in Mexico.

All books in this series on pioneers will be brief (up to about 62,000 words) and available as softcover and as eBooks. For libraries subscribing to SpringerLink, all readers of that institution will have free access to the electronic version and may also obtain a printed version as MyCopy. This makes these texts especially attractive for course adoption.

Each pioneer in this PSP series will be introduced by a colleague who is familiar with the pioneer's innovative work as a scholar or practitioner and who will place his or her work in the context of one or more scientific disciplines or political issue areas. This will be accompanied by an extended professional biography linking the selected pioneering academic's area with a personal biography and a comprehensive bibliography—or in the case of policymakers with a brief documentation of key innovative policy decisions. Then, the pioneer or a colleague will introduce selected benchmark papers, put them into the context of the author's own work, and reflect on the impact they have had on the continuing policy debate.

Mosbach, Germany, 14 May 2012 Hans Günter Brauch
Editor, SpringerBriefs on Pioneers in Science and Practice (PSP)
Editor, SpringerBriefs in Environment, Security, Development and Peace (ESDP)
Editor, Hexagon Series on Human and Environmental Security and Peace (HESP)

Preface

Arthur H. Westing has been known to me since the late 1970s or early 1980s when he was working at the *Stockholm International Peace Research Institute* (*SIPRI*), first as a Senior Researcher, and subsequently as Director of a project on the environmental impact of war that was later transferred to the *International Peace Research Institute Oslo* (*PRIO*). With the many books he authored and edited during this time, Westing became the single most important 'Pioneer on the environmental impact of war'. His pioneering scientific field work in Southeast Asia since the late 1960s on the use of defoliants by US and South Vietnamese forces during the Second Indochina War of 1961–1975 has influenced not only the scientific research and writing of many younger scholars in both peace research and environmental studies, but at the time also had significant impacts on the internal debate in the USA and increasing public opposition to that Viet Nam conflict.

The scientific work of Westing and several of his colleagues, to whom he refers elsewhere in this book (cf. Chap. 1), persuaded the US military to stop the use of Agent Orange; and it additionally influenced the US Administration in 1975 to submit, and the US Senate to ratify, the 1925 Geneva Protocol for the Prohibition of the Use in War of Asphyxiating, Poisonous or other Gases, and of Bacteriological Methods of Warfare (*LNTS*2138). Westing's work in the early and mid-1970s in the USA and during much of 1976–1990 in Stockholm and Oslo (during the *détente*period of East–West relations) also influenced the debates in the United Nations that contributed to the adoption of several international conventions and treaties, most particularly the 1972 Biological Weapon Convention (*UNTS*14860), the 1977 Environmental Modification Convention (*UNTS*17119), and the 1977 Geneva Protocol [I] on the Protection of Victims of International Armed Conflicts (*UNTS*17512). Indeed, Brauer (2009, p. 45) wrote in his *War and Nature* that the literature on the environmental effects of the Second Indochina War 'is, to a large extent, written or mediated by a single, formidable researcher, Arthur H. Westing.' He correctly concluded that, 'Westing may well be regarded as the father of the modern, continuous interest in the environmental effects of war'.

My own publications on chemical and biological warfare and on disarmament were stimulated by the many extremely innovative and valuable books and articles written by Westing and by the many personal discussions I had with him during the past three decades. As a biologist and forester, Westing has combined in his long and distinguished scientific career, his knowledge in forestry, botany, ecology, and conservation with the social responsibility of a scientist who—together Arthur W. Galston (Yale University), Egbert W. Pfeiffer (University of Montana), and Matthew S. Meselson (Harvard University)—produced the scientific evidence of the misuse of scientific knowledge in warfare. And it was Meselson and Westing who convinced the US Ambassador to Viet Nam, Ellsworth Bunker, to order the cessation of US herbicidal warfare when they informed him about their findings of the massive ecological and agricultural damage and possible human poisonings.

As a postdoctoral fellow myself at Harvard (1978) and as an active member of the Pugwash movement (1976–1992), I had met Meselson and many other natural scientists who contributed their scientific knowledge and concerns as citizens to constrain the misuse of their knowledge in warfare and to work for legally binding international arms control and disarmament treaties. Several scientists I met during the conferences and symposia organized by the Pugwash Movement considered it their patriotic duty during World War II and the subsequent Cold War to contribute their scientific knowledge—e.g., respectively, to the US, British, and Russian nuclear weapon programs—but once those had ended they founded and became involved with, e.g., the *Federation of American Scientists*, the *Union of Concerned Scientists*, and the *Society for Social Responsibility in Science* in order to get involved in the political debate on nuclear, biological, and chemical disarmament.

Westing's career has been different. He had never been involved in the development of weapons of mass destruction. His interest in plants, animals, and nature started when he was a boy scout and studied botany at Columbia University (B.A., 1950), when he interrupted his studies and became a Marine during the Korean War of 1950–1953, and later gained practical experience with the US Forest Service (1954–1955) where he conducted field research on means to kill unwanted hardwood trees through the use of herbicides that were later named Agent Orange by the military. As a Marine in Korea he gained field experience on the environmental disruption of warfare, and as a forester he gained practical experience in the use of herbicides, many years before *Silent Spring*, the seminal book by Carson (1962) was instrumental in initiating a global debate on the negative human interventions into nature, especially through the use of chemicals to optimize the economic output of agricultural, horticultural, and forest crops.

Westing's work as a forest biologist had in 1969 sensitized him to joining wildlife biologist Pfeiffer to verify onsite the increasing rumors of environmental devastation of forest and agricultural areas by aerially applied herbicidal anti-plant chemical warfare agents that was then being kept quite secret by the US government. At the invitation of the Government of Cambodia, Pfeiffer and Westing could visit attacked sites on the ground in December 1969 and January 1970. Based on those and 10 further field investigations to Southeast Asia, Westing and his colleagues (usually either Pfeiffer or Meselson) also examined and reported on crop destruction and the serious environmental damage from high-explosives. Their impact extended to the scientific community, to the US and other media, and to US Government and other political leaders (among them, Olof Palme) (cf. Chap. 1).

Based on this early scientific and field experience, Westing was invited by SIPRI to write a major report on his wartime findings that was published as *Ecological Consequences of the Second Indochina War* (Westing 1976) and subsequently to head a UNEP project on 'Peace, Security, and the Environment'. As a result of this long-term project at SIPRI and PRIO, Westing authored and edited a series of major monographs (cf. Chap. 3, #97; #108; #143; #145; #157; #163; #181; #193; #206; #218) in addition to various book chapters and articles (cf. Chap. 3, *passim*).

Westing's scientific work influenced the work of various intergovernmental agencies (e.g., UN, UNEP, UNIDIR, UNESCO); and it also raised the awareness of major nongovernmental organizations (e.g., the *International Committee of the Red Cross* [ICRC], the *International Union for Conservation of Nature* [IUCN], and the *International Campaign to Ban Landmines* [ICBL]).

As a professor of forestry, botany, ecology, and conservation at Purdue University (1959--1964), University of Massachusetts (1964–1965), Windham College (1966–1976), Hampshire College (1978–1983), and the European University Centre for Peace Studies (1989–2002) he introduced and inspired hundreds of students in several disciplines. When he was working at SIPRI and PRIO, Westing was a conceptual innovator in a number of scientific debates, especially on environmental, social, and comprehensive security, on the development of legal norms to stop the hostile use of herbicides, of hostile environment modification techniques, and of anti-personnel land mines and cluster bombs.

I have known Westing for more than 30 years as a highly respected, innovative, and critical natural scientist, environmentalist, and peace researcher who had a deep influence on my own work as a political scientist on weapon technologies, on the misuse of scientific knowledge for warfare, and on arms control and disarmament initiatives and policies.

As a scholar and concerned citizen, Westing through his scientific work and policy consultancy has lived what social responsibility of science stands for: i.e., to care about human beings and nature; to constrain military interference and damage to nature with herbicides having long-term effects on the food chain and thus not only on generations of children in the affected countries, but also on those of exposed soldiers. The affected individuals and their families have fought in vain for years to obtain medical assistance and financial compensation for their long-term health effects. One of the great successes of Westing and his colleagues was that they could convince Ambassador Bunker to stop the use of defoliants in the war in Southeast Asia. Nonetheless, when the USA in 1975 finally acceded to the 1925 Geneva Protocol, it did so with a reservation that would permit its military forces to use herbicides under certain conditions.

To summarize Westing's main environmental achievements during his professional life:

(a) As a *natural scientist* he has since the mid-1950s studied the use of defoliants in forestry, and through his 11 field trips to Southeast Asia between 1969 and 1993, he and his colleagues Pfeiffer and Meselson developed the first scientific evidence on the negative impact of the massive spraying of Agent Orange and other herbicidal anti-plant chemical warfare agents;
(b) As a *professor* and *educator* he has conveyed not only scientific knowledge, but also a sense of social responsibility to his students and colleagues, both nationally and internationally;
(c) As a *concerned US citizen* he and his colleagues had the courage to raise awareness both of the scientific community and the public at large about an essentially secret military operation of the USA and South Viet Nam during the Second Indochina War;
(d) As an *ecologist* and *environmentalist* he created awareness of the human interference into nature through the uncritical use of scientific knowledge as applied to warfare;
(e) As a *peace researcher* he put the environmental impacts of war on the research agenda of both critical peace and environmental studies and thus created a new field of multidisciplinary scientific endeavor;
(f) As a *policy consultant* he succeeded in putting the environmental impacts of warfare on the agenda of many intergovernmental agencies and nongovernmental organizations; and
(g) As a *politically active scientist and citizen* he became a *Vorbild* (model or example) not only for many young scientists, but also for policy makers, to take the courage not to remain silent on the misuse of scientific knowledge in warfare or for increasing short-term economic benefits while ignoring the longer term effects on the life of present and future generations.

In following Westing's work for more than three decades now, he has impressed me deeply though his modesty, his personal integrity, his social responsibility as a scientist, and his creating of public awareness. Westing has had an impact on the policies of his country, on the evolution of international legal norms, and on sensitizing future generations of scholars. For all these reasons I am both pleased and proud that, together with Westing, I could develop this new series on *Pioneers in Science and Practice*.

Hans Günter Brauch

References

Brauer, J. (2009). *War and nature: The environmental consequences of war in a globalized world* (233 pp). Lanham: AltaMira Press.

Carson, R. (1962). *Silent Spring* (368 pp). Boston: Houghton Mifflin.

Westing, A. H. (1976). *Ecological Consequences of the Second Indochina War* (119 pp + 8 plates). Stockholm: Almqvist & Wiksell.

Contents

Part I
War and the Environment

Chapter 1
The Environmental Impact of War: A Personal Retrospective

1.1 The Larger Context

Although past wars have, of course, been destructive of the environment to some greater or lesser extent (indeed, the same can be said for all wars), it was the intentional widespread, long-term, and severe destruction of the rural reaches of Indochina that contributed so poignantly to its worldwide notoriety once that US strategy became known.[1] It is abundantly clear that such wartime atrocities can arouse public opinion to the extent that they become the impetus for the adoption of new legal structures reflecting those expansions of public morality. Thus, by way of example, the extensive use of anti-personnel chemical warfare agents by the several major powers during World War I led to the widespread adoption of the 1925 Geneva *Protocol for the Prohibition of the Use in War of Asphyxiating, Poisonous or other Gases, and of Bacteriological Methods of Warfare* (LNTS 2138); and the attempted extermination of Jews and Gypsies by Germany during World War II led to the widespread adoption of the 1948 *Convention on the Crime of Genocide* (UNTS 1021). And it might be useful to note that such legal constraints are not only proscriptive, but normative as well.

And now we have a third example of a newly developed and widely accepted expansion of our cultural norms together with its translation into a legal norm—namely, that the several means of environmental destruction carried out by the USA during the Second Indochina War of 1961–1975, once they became known, led to the widespread adoption of the 1977 *Protocol [I] on the Protection of*

[1] The numbered references are provided in Chap. 3.

A. H. Westing, *Arthur H. Westing*, SpringerBriefs on Pioneers in Science and Practice 1, DOI: 10.1007/978-3-642-31322-6_1,

Victims of International Armed Conflicts (UNTS 17512) with its first of a kind inclusion of relevant environmental constraints.[2]

But as suggested below, that anti-environmental strategy was an unheralded one, and ultimately became generally known largely through the efforts of a small number of concerned natural scientists, I fortunately by chance among them. And it was a fortuitous combination of military experience, forestry training, and concern for the environment that made it possible for me to contribute usefully to this effort.

1.2 My Relevant Background

My interest in plants, animals, and the rest of nature began early in life and continues to this day. In my teens I was active in the outdoor activities of scouting (ultimately attaining the rank of Eagle Scout). As an undergraduate, I majored in botany (Columbia University, B.A, 1950), and then went on to graduate school to study forest ecology, silviculture, and tree physiology (Yale University, M.F., 1954; Ph.D., 1959). Toward the end of my two and a half years of service in the US Marine Corps (advancing from Second Lieutenant to Captain), I served for a time as the Forestry Officer of the Camp Lejeune military base. Importantly, during 1954–1955 I was employed by the US Forest Service to carry out a year of field research on means to kill unwanted hardwood (dicotyledonous) trees through the use, among others, of herbicides later code-named 'Agent Orange' by the US military (*e.g.*, #2). I subsequently spent some 24 years (1959–1983) in academia, variously professing forestry, botany, ecology, and conservation. Returning to my time in the Marine Corps, having received orders 'for service beyond the seas' in Korea, I spent a continuous year (1951–1952) in combat operations as an artillery officer (for most of that time in the front lines as a Forward Observer)—and thus not merely a witness to, but myself calling in the artillery fire responsible for large amounts of environmental disruption that I was later to come to regret (cf. Fig. 1.1).

1.3 The Second Indochina War ('Vietnam Conflict') of 1961–1975

The 1960s and 1970s were witness to the burgeoning of a widespread concern for the environment. A far-flung dislike of the US war in Indochina, although unrelated to those environmental concerns, intensified as the war progressed, both

---

[2] The Second Indochina War of 1961–1975 also led to adoption of the 1977 *Convention on the Prohibition of Military or any other Hostile Use of Environmental Modification Techniques* (*UNTS* 17119), unfortunately a relatively ineffectual treaty (#234).

Fig. 1.1 Arthur H. Westing, February 2011

within the USA and beyond. That antipathy was in time to be further compounded as news began to quietly filter out in the mid to late 1960s of the unheralded US strategy of denying forest concealment and sources of food to its elusive enemy through the use of herbicidal anti-plant chemical warfare agents. A small number of natural scientists in the USA, among them especially A. W. Galston of Yale

University and E. W. Pfeiffer of the University of Montana, soon became concerned over what was felt by them to be the misuse and corruption of socially beneficial scientific advances, ones to which Galston had even materially contributed.

In 1969 I had just been granted a sabbatical leave to continue my prior investigations into the morphological and physiological effects of gravity on conifers when the opportunity arose for me as a forest ecologist to join wildlife biologist, Pfeiffer to verify on-site the increasing rumors of environmental devastation of forest and agricultural areas by aerially applied herbicidal anti-plant chemical warfare agents. Earlier that year, Pfeiffer and another biologist had, in fact, visited South Viet Nam for that purpose, but at that time those assaults were still being kept more or less secret by the US Government, certainly the extent of them, so the two were essentially prevented from making on-site visits. Then when about a dozen of those herbicidal attacks were extended by the USA across the border into neutral Cambodia, Pfeiffer seized upon that opportunity for the two of us (plus two French colleagues) to visit those attacked sites. We were able to do this for several weeks in December 1969–January 1970 at the invitation of the Government of Cambodia and under the sponsorship of the *Scientists' Institute for Public Information* (#68). That inspection was followed by six others (together with either Pfeiffer or molecular biologist M.S. Meselson of Harvard University) to either South or North Viet Nam during the years 1970 through 1983, variously on behalf of the *Scientists' Institute for Public Information*, the *American Association for the Advancement of Science (AAAS)* (#79), the *Society for Social Responsibility in Science*, and (and during 1980–1983) the Government of Viet Nam. And during Pfeiffer's and my on-site investigations of the chemical forest destruction primarily via aerial spraying, we further recognized, examined, and reported on the then even less known crop destruction, also via aerial spraying (#67, #111); as well as on the awesome levels of previously unappreciated environmental destruction being brought about by high-explosive munitions (#63, #71, #95, pp 12-23, #149) and huge tractors (#57, #95, pp 46-50, #62).

Ellsworth Bunker was the US Ambassador to Saigon during 1967–1973, thus in essence the colonial viceroy of South Viet Nam. He considered the US goal to crush the Viet Cong and their North Vietnamese allies—and thereby to prevent South Viet Nam from falling to the Communists—to be of overriding importance to the southeast Asian interests of the USA. He fully supported, and basically ran the interminable war there, even backing the military aerial and ground incursions by the USA into neutral Laos and Cambodia, with the result of inflicting chaos upon Laos and, in time, unimaginable calamity upon Cambodia. I happened to know Bunker as a near Vermont neighbor. Indeed, this stood me in good stead during my second scientific mission to Indochina, this one on behalf of *AAAS* and headed by Meselson. I was able to meet with the Ambassador in Saigon in August 1970 (apparently as the first non-official visitor during his incredibly demanding and sensitive assignment), at which time he was not only willing and able to overcome the South Vietnamese Government's obstruction of the investigations of our group, but even arranged to place a helicopter at our disposal for almost two weeks, and asked us to report back to him at the end of our mission. To Bunker's enormous credit, when Meselson and I described our findings of massive ecological and agricultural damage and possible human poisonings, that led to his ordering a rapid end to the US herbicidal chemical warfare. He may have been in favor of the war, which he justified as a necessary crusade against communism, but at least he was not in favor of pursuing it by such flagrantly anti-environmental and anti-social means.

Our findings were widely aired during those years in the USA and elsewhere. Thus, I was invited to testify before the US Congress on four occasions: to the Senate Committees on Foreign Relations, on Commerce, and on the Judiciary, and to the US House of Representatives Committee on Science and Technology; and to lecture at the *US Arms Control and Disarmament Agency*. I spoke three times upon invitation to the US military: at the US Army Biological Laboratories at Fort Detrick; at the National War College in Washington; and at the US Military Academy at West Point, NY. In Saigon (now Ho Chi Minh City) I also reported to the South Vietnamese Ministry of Agriculture; and I was an expert witness at the Federal District Court of Minneapolis. I also made numerous public presentations, important among them: at the 1970 annual meeting of the *AAAS* in Chicago, doing so in my capacity as Director of its 'Herbicide Assessment Commission' (#79); at the *Centre International pour la Dénonciation des Crimes de Guerre* in Paris, as a private citizen; at the United Nations Conference on the Human Environment that was held in Stockholm in 1972, there representing the *Scientists' Institute for Public Information* (together with Pfeiffer) as well as the *Fellowship of Reconciliation*; at various fora in Sweden on behalf of the *Miljö Centrum*; and at the Twelfth International Botanical Congress, held in Leningrad in 1975, there as Professor of Ecology at Windham College.

It will be of interest to note that Pfeiffer and I were privileged to have a private session with Olof Palme, the Swedish Prime Minister, two days before he was to open the 1972 United Nations Conference on the Human Environment. We spent about an hour explaining our findings in words and pictures. Palme appeared to be very moved and asked many questions. It must be noted that during the pre-conference preparatory sessions, the USA had made it clear that it would boycott the event if military disruption of the environment were included in its agenda, a requirement that was, in fact, strictly adhered to. However, when Palme opened the Conference with his welcoming address he added several long minutes to his prepared speech decrying anti-environmental tactics by armed forces in an already environmentally beleaguered planet. Although Palme had not mentioned the United States in those added remarks, there was no doubt as to what he was referring. This jab so infuriated the USA that it immediately recalled its Ambassador from Sweden for 'consultations'.

In the end, my on-site investigations and related activities took me to Indochina on 11 occasions during and following the war (in 1969–1970, 1970, 1971, 1973, 1980, 1982, 1983, 1991, 1992, 1992, and 1993), each time for two or more weeks on the ground there.

Publications of mine on the military assault on the environment appeared in US and international scholarly journals and newspapers, starting in early 1970 with a paper entitled 'Poisoning plants for peace' (#50), and soon to be followed by numerous others (*e.g.*, #54, #56, #61, #64, #74, #80, #86, #87, #98, #101, #112), including the one reproduced elsewhere in this book (#91, cf. Chap. 4). Two especially pernicious societal concerns arising from the US actions in Indochina that took some of my attention were: first, the widely distributed *dioxin* as an inadvertent toxic contaminant in the most heavily employed anti-plant chemical warfare agent (#99, #276); and, second, the long-lasting residue of the huge amounts of *unexploded ordnance*, the so-called explosive remnants of war, including especially anti-personnel land mines and cluster bomb units (#93, #138, #141, #284, #295).

1.4 And Following the War in Indochina

The activities summarized above resulted in my being asked by the *Stockholm International Peace Research Institute* (SIPRI) to prepare a major report on my wartime findings, which resulted in the 1976 *Ecological Consequences of the Second Indochina War* (#95). That invitation led in turn to my being asked by SIPRI to direct a project being farmed out to it by the *United Nations* on 'Peace, Security, and the Environment', which was a component of its newly established 'System-wide Medium-term Environment Progamme' under the charge of the *United Nations Environment Programme* (UNEP) (#408).[3] As a result of accepting that offer, my combined times in residence at SIPRI added up to about eight years during the period 1975–1987, plus almost a further three years (1988–1990) at the *Peace Research Institute Oslo* (PRIO) when the project was transferred to there. And it further led to war-related consultancies both during and after the war with, among others: (**a**) the *United Nations Institute for the Disarmament Research* (UNIDIR) (cf., e.g., #218, #233); (**b**) the *International Committee of the Red Cross* (ICRC) (cf., e.g., #284); (**c**) the *United Nations Scientific, Educational and Cultural Organization* (UNESCO) (cf., e.g., #380, #393); and (**d**) UNEP (cf., e.g., #208, #247—and importantly including an analysis of the environmental impact of the Gulf War of 1991, *i.e.*, #209, #389, cf. Chap. 5).

Following my investigations of the environmental impact of the Indochina war, my efforts within the SIPRI/PRIO/UNEP project progressed quite naturally toward considerations of how such future depredations might be curtailed, whether they be from conventional, chemical, biological, or nuclear war. Those inquiries perforce involved both legal and cultural norms. For the legal norms (a subject in

---

[3] It is a pleasure for me to note that UNEP has recently returned to the issue of war *vis-à-vis* the environment. For the inaugural report in this revived policy series, see: *From Conflict to Peacebuilding: the Role of Natural Resources and the Environment* by R. A. Matthew, O. Brown, & D. Jensen (Nairobi: *United Nations Environment Programme* Policy Paper No. 1, 44 pp, 2009).

which I had had no prior experience) it was my good fortune to have as a SIPRI colleague, Jozef Goldblat, a preeminent authority on the Law of War (International Humanitarian Law, including International Arms Control and Disarmament Law). And thinking about cultural norms (also then new to me) led me to the broader closely related issue of what actually constitutes environmental security, whether at a national, regional, or global level—and approaches toward achieving it in a consistently bellicose (and increasingly overpopulated and over-consumptive) world. I come back to the implications of this further on in the book (cf. Chap. 2).

Chemical, biological, and nuclear weapons—generally referred to as 'unconventional' weapons—have the potential for being variously deadly and destructive of humans, of infrastructure, and of the environment. (The distinction between anti-personnel and anti-plant chemical weapons is to some extent technically feasible, but not at all diplomatically.) Fortunately there exists a widespread public antipathy to the use of all such unconventional weapons. And when chemical weapons have been employed in the recent past this has usually been done without acknowledgement. On the other hand, they do all exist and could thus be used in the future. I discuss the impact of hostile manipulations of the environment (so-called environmental warfare) elsewhere in the book (cf. Chap. 6), and also the impact of nuclear weapons (cf. Chap. 7), but can refer the reader to other of my writings that examine chemical weapons (#151, #154, #208), biological weapons (#158, #177, #212), or the two together (#155, #186, #208).

So thus in sum, as noted above, I have shared with others my findings and thoughts on the nexus between environment and war through numerous publications (cf. Chap. 3). Of the books among those publications: I was the sole author of three on behalf of SIPRI alone (#95, #97, #108); the editor and co-author of five on behalf of SIPRI plus UNEP (#143, #145, #157, #163, #181); the editor and co-author of two on behalf of PRIO plus UNEP (#193, #206); the editor and co-author of one on behalf of UNIDIR plus UNEP (#218); and the editor and co-author of one on behalf of UNEP alone (#247)—the last two of these as a private consultant.[4]

1.5 Related Developments

The question arises of what my efforts, together with those of my colleagues, have actually led to—not an easy question to answer. To begin with, as already noted earlier, Meselson's and my presentation to Ambassador Bunker led to the rapid cessation by the USA of the hostile employment in Indochina of chemical warfare agents. Moreover, the publicity we helped to generate about their use was certainly

[4] In 1990, Carol E. Westing (my wife) and I established Westing Associates in Environment, Security & Education (Address: 134 Fred Houghton Rd, Putney, VT 05346, USA. westing@sover.net).

instrumental in finally, in 1975, having the USA accede to the 1925 Geneva *Protocol for the Prohibition of the Use in War of Asphyxiating, Poisonous or other Gases, and of Bacteriological Methods of Warfare* (LNTS 2138).

The results of our studies also contributed materially to the formulation of two further treaties, viz., the 1977 *Protocol [I] on the Protection of Victims of International Armed Conflicts*, especially its Articles 35.3 and 55 (UNTS 17512) and the 1977 *Convention on the Prohibition of Military or any other Hostile Use of Environmental Modification Techniques* (*UNTS* 17119). Similarly, my studies (and consultancy to the *International Committee of the Red Cross* in Geneva) on the pernicious aspects of anti-personnel mines and other unexploded ordnance in time contributed in at least some small way to the acceptance of two further treaties, viz., the 1997 Anti-personnel Land Mine Convention (UNTS 35597) and the 2008 Cluster Munition Convention (UNTS 47663).

More generally, my efforts helped to establish *environmental disruption by military actions* as a distinct (and now ever growing) area of study by scholars in the fields of both political science (especially in peace research) and natural science (especially in environmental studies).[5] I believe that these efforts have also sensitized military authorities to the importance of minimizing environmental battlefield disruption. And I was among those early on who developed the parameters of environmental security (cf. Chap. 2), now a widely accepted concept.

1.6 A Final Note of Appreciation

Of the various people to whom I owe debts of gratitude, a number stand out with particular clarity: (a) It was *Arthur W. Galston* (1920–2008), my major doctoral professor and lifelong friend and mentor, who started me on the path that is outlined in these pages. (b) It was *E.W. (Bert) Pfeiffer* (1915–2004), fellow ex-Marine, fellow natural scientist, and close friend, with whom I was foolhardy enough to brave and study the active war zones of Indochina. (c) It was *Eugene C. Winslow*, my President at Windham College, fellow natural scientist, and supporter

[5] Two recent books that very generously attest to my contributions in these fields are: (a) *War and Nature: the Environmental Consequences of War in a Globalized World* by Jurgen Brauer (Lanham, MD, USA: AltaMira Press, 233 pp. 2009); and (b) *The Invention of Ecocide: Agent Orange, Vietnam, and the Scientists who Changed the Way We Think about the Environment* by David Zierler (Athens, GA, USA: University of Georgia Press, 245 pp. 2011). Additionally I should note that the importance of these groundbreaking studies of the environmental consequences of warfare has been recognized by an honorary doctorate (DSc, Windham, 1973) as well as by medals from both the New York Academy of Sciences (1983) and Government of Bulgaria (1984); and also by being named a 'Peace Messenger' (together with four international colleagues) by the United Nations Secretary-General (1987) and by being selected as one of the 500 individuals worldwide to have been appointed to the United Nations 'Global 500 Roll of Honour' (1990).

of my efforts, who went out of his way to grant my several leaves of absence during those Indochina War years; and it was Adele Smith Simmons, my President at Hampshire College, who was similarly helpful in the postwar years. (d) It was *Frank Barnaby*, natural scientist and Director of *SIPRI*, *Sverre Lodgaard*, political scientist and Director of *PRIO*, *Naigzy Gebremedhin*, land planner and Head of Technology at *UNEP*, and *Mostafa K. Tolba,* natural scientist and Executive Director of *UNEP*, who (despite objections from the USA) all so unstintingly and liberally supported my *SIPRI/PRIO/UNEP* program and validated its results. (e) It is Jeanne and Stephen, my two then young children, who seemed to tolerate and perhaps even forgive those absences. (f) It is *Hans Günter Brauch*, long-time colleague and fellow peace researcher, for his unexpected and heart-warming initiative to have Springer Verlag honor me with the designation of *Pioneer on the Environmental Impact of War*. And finally, (g) It is *Carol Eck Westing*, wife, close companion, gentle and insightful critic, to whom I owe the largest debt of gratitude for tolerating my eventful absences and, of course, for not standing in the way of my pursuit of these endeavors.

Chapter 2
What Next? A Search for Security in War and Peace

2.1 As Matters Now Stand

It has become abundantly clear that the global biosphere—its plants, animals, and associated ecosystems (biomes)—are being ever more seriously threatened by ever greater human arrogations on the one hand, and by ever greater human disruptions on the other. As our human numbers keep increasing, our combined human needs perforce increase apace; and, in turn, those needs are compounded by our even more rapidly growing discretionary uses.[1] One poignant example of our global over-population is the vast number of environmental refugees. The ongoing environmental devastation is, of course, a tragedy in its own right, but that environmental devastation impinges as well on what I have defined as 'comprehensive human security' with its unavoidable *environmental security* component.

Our inexcusably unsustainable utilization of the world's renewable natural resources (woodland, grassland, fresh water, and ocean over-extractions)—as well as our inexcusably unsustainable utilization of the world's sink capacities (terrestrial, marine, and atmospheric) via dumpings of solid, liquid, and gaseous wastes into all of those domains—derive largely, of course, from the now increasingly over-populated and over-consumptive *civil sector* of society. However, society's *military sector* adds a certain amount to those environmental assaults from its peacetime activities (#123, #146, #182), and a substantial—and potentially huge—amount from its wartime activities. It has thus been my hope that at least the military contributions to these shortsighted actions can be minimized, especially those in wartime that lead to widespread long-term, and severe damage to the environment.

The most straightforward and elegant approach to dealing with the issue of wartime environmental disruption would, of course, be to eliminate an armed force's recourse to achieving its aims through armed conflict, whether that force be

[1] The numbered references are provided in Chap. 3.

A. H. Westing, *Arthur H. Westing*, SpringerBriefs on Pioneers in Science and Practice 1, DOI: 10.1007/978-3-642-31322-6_2,

governmental or rebel (guerilla, insurgent) (#202). The problem here, to lean on the 1945 *UNESCO* Constitution (UNTS 52), is that 'since wars begin in the minds of men, it is in the minds of men that the defences of peace must be constructed'. Those greatly needed peace-establishing 'constructions' have withstood the efforts of so many throughout human history as to make it clear to me at least that armed conflict has been, and will continue to be, a characteristic human endeavor. Indeed, there has probably never been a day of peace among humans throughout the long sweep of our existence on earth; and in modern times there are always a dozen or more armed conflicts in progress somewhere or other, whether international or non-international (#127, #179, pp 3–4). So the hope must be to minimize the deadly and destructive impacts of our many, many wars, doing so for both eco-centric and anthropocentric reasons. As outlined next, the approach in my view will have to be largely through the Law of War as reinforced, and in time strengthened, by a more socially and environmentally sensitized public opinion, that is to say, by appropriately expanded cultural norms (#365). And, more generally, those expanded cultural norms would serve as a prerequisite for strengthening environmentally and socially sensitive global governance.

2.2 The Law of War

Before the anti-environmental aspects of the Second Indochina War of 1961–1975 (cf. Chap. 4) and the Gulf War of 1991 (cf. Chap. 5) fade from the collective memory of the public at large, efforts must be made not only to keep that memory alive, but as well to ensure that the constraints on environmental destruction now imbedded in the Law of War (in International Humanitarian Law) are incorporated into the military manuals and rules of engagement of all the world's armed forces (cf. Chap. 9). And it then becomes crucial that those existing constraints be widely publicized and, moreover, emphasized especially during officer training. Those Governments whose armed forces do not as yet have such documents should be convinced to develop them (perhaps with the readily available expert assistance of the *International Committee of the Red Cross* in Geneva).

Since so many of today's wars are of a non-international (internal) nature, or largely so, the important question arises of how to instill the appropriate environmental norms into the many armed forces not under their Government's control. Most of the Law of War is formally applicable only to international armed conflicts, and perhaps the primary multilateral instrument of relevance to internal wars, 1977 *Protocol [II] on Non-international Armed Conflicts* (UNTS 17513), provides only rather weak strictures. Moreover, Governments consider such armed conflicts to be internal matters and therefore not open to outside interference. Thus, as politically and diplomatically sensitive as the matter is, it might well be useful to have outside groups suggesting to rebel forces that for them to openly adhere to the Law of War would not only be beneficial to the land and people of their own country and thereby give strength to their cause, but could as well help to legitimize it to the outside world.

Self-inflicted environmental damage in wartime, a self-inflicted so-called 'scorched-earth' tactic, occurs reasonably often. The Law of War is silent on this matter, and considerations of whether it is possible to minimize such damage is not here considered further, except indirectly insofar as widespread public education might be of some help to curb such instances.

The Law of War does now incorporate a certain amount of environmental protection from military actions (#154, #179, #232; #311, cf. Chap. 8). The important question arises of whether efforts should be made to strengthen the current Law of War with further constraints on environmental disruption. Here I would suggest that the now existing constraints are about as restrictive as most Governments will currently accept (with some of the major powers considering even the existing ones too onerous to accede to). Widespread *environmental education*—both *formal* at all levels of schooling, and *informal* especially abetted by the efforts of inter-governmental agencies and non-governmental organizations—is of the utmost importance (#241). Short of some future environmental cataclysm, it will have to be through such multifaceted environmental education that there can be any hope for the needed change to occur in public attitudes (societal values, cultural norms). Such change could on the one hand produce the necessary impetus for Governments to adopt new multilateral instruments strengthening the existing constraints on ecocidal actions (indeed, whether military or civil), and on the other to have them in fact be adhered to.

2.3 The Search for Security

So our mission for the future will have to be to strive to come ever closer to attaining environmental security together with social (societal) security—that is, to attempt this within the framework of comprehensive human security (#162, #188, #196, #210, #237, #371), a concept fully applicable to both non-industrialized (#213, #244, #277) and industrialized states (#294). There is, in fact, some modest hope for the viability of that aim since our cultural norms—first their social components, and more latterly their environmental components—have been to some extent evolving in the right direction since at least the end of World War II (#380).

As I have thought of it, *comprehensive human security* consists of a number of inexorably intertwined *environmental* and *social* components, with neither of those two categories attainable in the long run unless both are. And, additionally, neither of which are in the long run attainable unless human numbers become compatible with available necessities. Thus, to summarize the make-up of the two components of comprehensive human security: *Environmental security* is comprised of two sub-components: (a) *rational resource utilization*, that based on use or harvesting at levels and with procedures that maintain or restore optimal resource services or stocks; and (b) *environmental protection*, that based on protection from at least medically unacceptable pollution, protection from permanent human intrusions in special areas (comprising at least 15 % of the global biosphere)[2]; and, of course, protection from avoidable disruptive military actions. And *social security* is comprised of four sub-components: (a) *political safeguards*, those based on participatory democracy by an informed public, a free press, and a robust legal system; (b) *economic safeguards*, those based on a guaranteed minimum income, access to housing, medical care, care of the aged, child care, and education; (c) *personal safeguards*, those based on justice, equity, equality of the sexes, and respect for others; and, of course, (d) *military safeguards*, those based on a purely defensive, non-provocative posture, and the rejection of weapons of mass destruction.

And, I must emphasize, neither environmental security nor social security will ever be realized: (a) unless there is widespread transfrontier (*i.e.*, regional) cooperation, for the simple reason that ecosystem boundaries rarely coincide with political boundaries (#191, #236, #245, #271, #287, #317, #318, #328); and also, for that matter, because few states are self-sufficient as to needed natural and other resources; and (b) unless human numbers overall become compatible with available necessities (#116, #203).

The task before us is to insure the worldwide pursuit of pervasive education, both formal and informal—in simplest terms, for the purpose of instilling the notions embodied in the 1948 *Universal Declaration of Human Rights* (*United Nations General Assembly* Resolution No. 217[III] A, 10 Dec 48) on the one hand,[3] and in the 1982 *World Charter for Nature* (*United Nations General Assembly* Resolution No. 37/7, 28 Oct 82) on the other. The widespread knowledge, understanding, and acceptance of the fundamentally important concepts enunciated by those two benchmark documents will in due course go a long way toward reducing environmental damage of the earth—whether of military or civil origin—an earth upon which all of us unavoidably depend, and upon which the other creatures with which we perforce share this earth also unavoidably depend.

So my hope is that pressure from an informed and sensitized public will lead to the necessary reorientation and restructuring of national priorities throughout the world in order to achieve the inexorably intertwined national environmental and social securities outlined above. But national restructuring will certainly not suffice without greater regional and global cooperation, to be achieved through a concomitant restructuring and strengthening of regional and global governance

---

[2] In fact, in 2010 the states parties to the 1992 *Convention on Biological Diversity* (UNTS 30619) concluded that it was necessary to conserve at least 17 % of the world's terrestrial and inland water areas and 10 % of its coastal and marine areas (www.cbd.int/sp/targets).

[3] Most of the aspirational 1948 *Universal Declaration of Human Rights* (*United Nations General Assembly* Resolution No. 217[III] A, 10 Dec 48) was subsequently formalized via a pair of complementary multilateral treaties: the 1966 *International Covenant on Economic, Social and Cultural Rights* (UNTS 14531); and the 1966 *International Covenant on Civil and Political Rights* (UNTS 14668).

systems. And the needed strengthening of global governance will in turn have to be realized through the widespread acceptance of a more powerful United Nations system, with UNEP serving as one of the key actors in that system, working in cooperation with an upgraded *United Nations Commission on Sustainable Development* (UNCSD).

Chapter 3
The Author's Relevant Papers: A Selective Listing

Note: *The author of all the following entries is 'Westing, Arthur H.', these having been extracted from my sequential life list of publications. The number preceding each title refers to its sequential number in that compilation. Publications by me to which reference is made elsewhere in this text are keyed to that number. The six publications presented* in toto *in Chapters 4 through 9 (and republished by permission of the copyright holders) are noted as *.*

#2 Effects of undiluted 2,4-D and 2,4,5-T in cut surfaces on oak in Lower Michigan. *Annual Research Report of the North Central Weed Control Conference* (Champaign, IL, USA) 12:168. 1955.

#50 Poisoning plants for peace. *Friends Journal* (Philadelphia) 16(7):193–194. 1 Apr 1970.

Reprinted in: *Vermont Freeman* (Plainfield, VT, USA) 2(4):7–9. 30 Jan–1 Feb 1970.
Reprinted in: Weisberg, B. (ed.). *Ecocide in Indochina: the Ecology of War.* San Francisco: Canfield Press, 241 pp: pp 72–74. 1970.
Reprinted in: Glad Day Press (ed.). *Not since the Romans Salted the Land: Chemical Warfare in S.E. Asia.* Ithaca, NY, USA: Glad Day Press, 28 pp: pp 16–17. 15 May 1970.

#54 Ecocide in Indochina. *Natural History* (New York) 80(3):56–61,88. Mar 1971.

Reprinted in: *Brattleboro [VT, USA] Daily Reformer* 58(287):5. 5 Feb 1971.
Reprinted in: *Vermont Freeman* (Plainfield, VT, USA) 3(7):4. 1 Mar 1971.
Reprinted in: *US Congressional Record* (Washington) 117(6): 8183–8184. 25 Mar 1971.

A. H. Westing, *Arthur H. Westing*, SpringerBriefs on Pioneers in Science and Practice 1, DOI: 10.1007/978-3-642-31322-6_3,

Reprinted in: Moynihan, W.T. (ed.). *Essays Today*. New York: Harcourt Brace Jovanovich, 224 pp: pp 54–58. 1972.
Reprinted in: Ternes, A. (ed.). *Ants, Indians, and Little Dinosaurs*. New York: Charles Scribner's Sons, 391 pp: pp 292–297. 1975.

#56 Ecological effects of military defoliation on the forests of South Vietnam. *BioScience* (Washington) 21(17):893–898,889. 1 Sep 1971.

Reprinted in Japanese in: *Seibutsu Kagaku* (Tokyo) 23(3):161–168. Mar 1972.

#57 The wasteland: beating plowshares into swords (with Haseltine, W.). *New Republic* (Washington) 165(18):13–15. 30 Oct 1971.

#61 Forestry and the war in South Vietnam. *Journal of Forestry* (Washington) 69(11):777–783. Nov 1971. (Cf. *ibid.* 70(3):129. Mar 1972.)

#62 Leveling the jungle. *Environment* (Washington) 13(9):8–12. Nov 1971.

Reprinted in: *Vermont Freeman* (Plainfield, VT, USA) 3(19):18–19. Early Oct 1971.
Reprinted in: *Indochina Chronicle* (Berkeley, CA, USA) 1972(13):6–10. Jan 1972.
Reprinted in French in part in: *Le Courrier du Vietnam* (Hanoi) 9(360): 8. 24 Feb 1972.
Reprinted in: *US Congressional Record* (Washington) 118(5): 6478–6479. 1 Mar 1972.
Reprinted in: *US Congressional Record* (Washington) 118(6): 6876–6877. 3 Mar 1972.
Reprinted in part in: *American Report* (New York) 2(12):9. 3 Mar 1972.
Reprinted in part in: *Quaker Service Bulletin* (Philadelphia) 1972(112):5. Spr 1972.
Reprinted in Finnish in: *Ydin* (Helsinki) 6(6):14–15. Jun 1972.
Reprinted in: Pell, C. (ed.). *Prohibiting Military Weather Modification*. Washington: US Senate Committee on Foreign Relations, 162 pp: pp 119–122. 1972.

#63 The big bomb. *Environment* (Washington) 13(9):13–15. Nov 1971. (Cf. *ibid.* 14(2):48. Mar 1972 -&- *ibid.* 14(8):43. Oct 1972.)

Reprinted in: *Vermont Freeman* (Plainfield, VT, USA) 3(18): 18–19;(20):22. Sep-Latter Oct 1971.
Reprinted in part in: *Fellowship* (Nyack, NY, USA) 37(11):3–4. Nov 1971.
Reprinted in: *Indochina Chronicle* (Berkeley, CA, USA) 1972(13): 11–13. Jan 1972.
Reprinted in French in part in: *Le Courrier du Vietnam* (Hanoi) 9(360): 7–8. 24 Feb 1972.

Reprinted in: *US Congressional Record* (Washington) 118(5): 6479–6480. 1 Mar 1972.
Reprinted in: *US Congressional Record* (Washington) 118(6): 6877–6878. 3 Mar 1972.
Reprinted in part in: *American Report* (New York) 2(21):9. 3 Mar 1972. (Cf. ibid. 2(45):3. 8 Sep 1972.
Reprinted in Finnish in: *Ydin* (Helsinki) 6(6):15. Jun 1972.
Reprinted in: Pell, C. (ed.). *Prohibiting Military Weather Modification*. Washington: US Senate Committee on Foreign Relations, 162 pp: pp 122–124. 1972.

#64 Herbicides as agents of chemical warfare: their impact in relation to the Geneva Protocol of 1925. *Environmental Affairs* (Hanover, NH, USA) 1(3):578–586. Nov 1971 (1971–1972).

Reprinted in: Fulbright, J.W. (ed.). *Geneva Protocol of 1925*. Washington: US Senate Committee on Foreign Relations, 439 pp: pp 234–247. 1972.

#67 The U.S. food destruction program in South Vietnam. In: Browning, F., & Forman, D. (eds). *Wasted Nations*. New York: Harper & Row, 346 pp: pp 21–25. 1972.

Earlier version in: *New York Times* 120(41,442):27. 12 Jul 1971.
Earlier version in: *Vermont Freeman* (Plainfield, VT, USA) 3(8):9–10. 8 Mar 1971.
Earlier version in: *ADA World* (Washington) 26(7-8-9):2,14. Sep 1971.
Earlier version in: *Solidarity with Vietnam* (Hanoi) 1971(33):12–15. Oct 1971.

#68 Herbicidal damage to Cambodia. In: Neilands, J.B., *et al. Harvest of Death: Chemical Warfare in Vietnam and Cambodia*. New York: Free Press, 304 pp: pp 177–205 (Chap. 4). 1972.

#71 The cratering of Indochina (with Pfeiffer, E.W.). *Scientific American* (New York) 226(5):20–29,138,16;(6):7. May, Jun 1972. (Cf. *ibid.* 227(3):8,12. Sep 1972.)

Reprinted in: *Los Angeles Times*, 30 Apr 1972, pp G1,G7.
Reprinted in: *Boston Globe* 201(128):1,22. 7 May 1972.
Reprinted in part in: *Toronto Star*, 8 May 1972, p. 37.
Reprinted in Finnish in: *Ydin* (Helsinki) 6(6):12–14. Jun 1972.
Reprinted in part in: *Windham College Bulletin* (Putney, VT, USA) 22(2):2–3. Smr 1972.
Reprinted in: *Explorer* (Cleveland, OH, USA) 14(3):8–14. Fall 1972.
Reprinted in Italian in: *Paese Sera* (Rome) 23(281):3. 13 Oct 1972.
Summary in part in French in: *Jeune Afrique* (Paris) 1972(618):3. Nov 1972.

Reprinted in: York, H.F. (ed.). *Arms Control.* San Francisco: W.H. Freeman, 427 pp: pp 329–338,420. 1973.
Reprinted in: *Annals of America. XIX. 1969–1973: Deténte and Domestic Crisis.* Chicago: Encyclopædia Britannica, 439 pp: pp 286–290. 1974.

#74 Herbicides in war: current status and future doubt. *Biological Conservation* (Barking, UK) 4(5):322–327. Oct 1972 (1971–1972).

Similar in: *Vermont Freeman* (Plainfield, VT, USA) 3(20):18–19. Latter Oct 1971.
Similar in: *Science Today* (Bombay [now Mumbai]) 6(8):43–46. Feb 1972.

#79 AAAS Herbicide Assessment Commission. *Science* (Washington) 179(4080):1278–1279. 30 Mar 1973.

#80 Ecocide: our last gift to Indochina. *Environmental Quality Magazine* (N. Hollywood, CA, USA) 4(5):36–42,62,64–65. May 1973.

Reprinted in: *US Congressional Record* (Washington) 119(13): 16468–16470. 22 May 1973.

#86 Proscription of ecocide: arms control and the environment. *Bulletin of the Atomic Scientists* (Chicago) 30(1):24–27. Jan 1974.

Reprinted in: Falk, R.A. (ed.). *Vietnam War and International Law. IV. The Concluding Phase.* Princeton, NJ, USA: Princeton University Press, 1051 pp: pp 283–286. 1976.
Reprinted in part in: Thee, M. (ed.). *Armaments and Disarmament in the Nuclear Age: a Handbook.* Stockholm: Almqvist & Wiksell, 308 pp: pp 142–145. 1976. (Reprinted by various publishers in Finnish, French, German, Japanese, Norwegian, & Serbo-Croatian editions.)
Preliminary version in: *Vermont Freeman* (Plainfield, VT, USA) 4(20):6–9. Latter Oct 1972.

#87 Postwar forestry in North Vietnam. *Journal of Forestry* (Washington) 72(3):153–156. Mar 1974.

#91* Environmental consequences of the Second Indochina War: a case study. *Ambio* (Stockholm) 4(5–6):216–222. 1975.

Reprinted in Swedish in: *Vår Lösen* (Stockholm) 67(4–5):213–222. Apr–May 1976.
Reprinted in part in: *SIPRI Yearbook* (Oxford) 1976:82–83.
Reprinted in: *Strategic Digest* (New Delhi) 7(1–2):72–81. Jan–Feb 1977.
[***Nb:*** Reproduced in Chapter 4 by permission of the *Royal Swedish Academy of Sciences*, the copyright holder, on 21 March 2012.]

#93 The unexploded munitions problem: an American legacy to Indochina that still wounds and kills. *Sunday Rutland [VT, USA] Herald & Times Argus* 1(10)(Sect. 4):3. 14 Dec 1975.

Reprinted in part in: *Indochina Chronicle* (Berkeley, CA, USA) 1976(49):4. May–Jun 1976.

#95 *Ecological Consequences of the Second Indochina War.* Stockholm: Almqvist & Wiksell, 119 pp + 8 plates. 1976.

Reprinted in part in: *SIPRI Yearbook* (Oxford) 1976:48–53.
Synopsis in: *SIPRI Yearbook* (Oxford) 1977:198–200.
Reprinted in Japanese: Tokyo: Iwanami Shoten Publishers, 220 pp + 8 plates. 1979.

#97 *Weapons of Mass Destruction and the Environment.* London: Taylor & Francis, 95 pp. 1977.

#98 Ecological effects of the military use of herbicides. In: Perring, F.H., & Mellanby, K. (eds). *Ecological Effects of Pesticides.* London: Academic Press, 193 pp: pp 89–94. 1977.

#99 Ecological considerations regarding massive environmental contamination with 2,3,7,8-tetrachlorodibenzo-*para*-dioxin. In: Ramel, C. (ed.). *Chlorinated Phenoxy Acids and their Dioxins: Mode of Action, Health Risks and Environmental Effects.* Stockholm: Swedish Natural Science Research Council, Ecological Bulletin No. 27, 302 pp: pp 285–294. 1978.

Reprinted in part in: *SIPRI Yearbook* (Oxford) 1977:86–102 (Chap. 4).
Reprinted in part in: *Atlas World Press Review* (New York) 24(11):13–15. Nov 1977.

#101 The military impact on the human environment. *SIPRI Yearbook* (Oxford) 1978:43–68 (Chap. 3).

Reprinted in part in: *Scientific World* (London) 22(1):22–24. 1978. (Also in the French, German, & Russian [pp 29–32] editions.)

#102 Neutron bombs and the environment. *Ambio* (Stockholm) 7(3):93–97. 1978.

Reprinted in part in Swedish in: *Pax* (Stockholm) 1978(3):5–7. 1978.
Reprinted in German in: *Wissenschaft und Fortschritt* (Berlin) 28(12):446–448. Dec 1978.
Summary in: *New Scientist* (London) 79(1114):325. 3 Aug 1978.
Summary in: *Sweden Now* (Stockholm) 12(4):50–51. Jul–Aug 1978.
Summary in: *Scientific American* (New York) 239(4):85. Oct 1978.

#108 *Warfare in a Fragile World: Military Impact on the Human Environment.* London: Taylor & Francis, 249 pp. 1980.

Also available (15.8 MB, pdf) at: http://books.sipri.org/product_info?c_product_id=269

#111 Crop destruction as a means of war. *Bulletin of the Atomic Scientists* (Chicago) 37(2):38–42. Feb 1981. (Cf. *ibid.* 37(8):61. Oct 1981.)

#112 Endangered species and habitats of Viet Nam (with Westing, C.E.). *Environmental Conservation* (Cambridge, UK) 8(1):59–62. Spr 1981.

#116 A world in balance. *Environmental Conservation* (Cambridge, UK) 8(3):177–183. Aut 1981.

#119 Environmental impact of nuclear warfare. *Environmental Conservation* (Cambridge, UK) 8(4):269–273. Wntr 1981.

Reprinted in part in: *Ambio* (Stockholm) 11(2–3):147. 1982.

#122 The environmental aftermath of warfare in Viet Nam. *SIPRI Yearbook* (Oxford) 1982:363–389 (Chap. 11).

Reprinted in: *Natural Resources Journal* (Albuquerque, NM, USA) 23(2):365–389. Apr 1983.

#123 Environmental quality: the effect of military preparations. *Environment* (Washington) 24(4):2–3,39–40. May 1982.

Earlier version in: *New Perspectives* (Helsinki) 11(5):27–29. 1981.
Earlier version in: Prchlík, V. (ed.). *Peace, Energy and the Environment.* Prague: Czechoslovak Peace Committee, 148 pp: pp 55–65. 1981.

#127 War as a human endeavor: the high-fatality wars of the twentieth century. *Journal of Peace Research* (Oslo) 19(3):261–270. 1982.

#131 Environmental consequences of nuclear warfare. In: Gore Jr, A. (ed.). *Consequences of Nuclear War on the Global Environment.* Washington: US House of Representatives Committee on Science & Technology, 260 pp: pp 141–146. 1983.

Reprinted in: *Environmental Conservation* (Cambridge, UK) 9(4):269–272. Wntr 1982.

#134 The environmental imperative of nuclear disarmament (with Tolba, M.K., & Polunin, N.). *Environmental Conservation* (Cambridge, UK) 10(2):91–95. Smr 1983.

#138 Explosive remnants of conventional war (with Abdussalam, A.A., Anderberg, B., Goldblat, J., Gordon, G., Kassas, M., Molski, B.A., & Sgaier, K.). In: Pérez de Cuéllar, J. (ed.). *Problem of Remnants of War.* New York: *United Nations General Assembly* Document No. A/38/383, 28 pp: pp 6–28. 19 Oct 1983.

Reprinted in: Westing, A.H. (ed.). *Explosive Remnants of War: Mitigating the Environmental Effects*. London: Taylor & Francis, 141 pp: pp 117–136 (App. 8). 1985.
Reprinted in Finnish in: *Peace Union of Finland* (Helsinki) *Disarmament Pamphlet* No. 30, 32 pp. Mar 1989.

#139 How much damage can modern war create? In: Barnaby, F. (ed.). *Future War*. London: Michael Joseph, 192 pp: pp 114–124 (Chap. 10). 1984.

(Also published in a Spanish edition by Editorial Debate, Madrid. 1985.)

#140 Ydinsodan pitkäaikaisvaikutukset [Long-term effects of a nuclear war] (in Finnish). *Suomen Lääkärilehti* [*Finnish Medical Journal*] (Helsinki) 39(4):231–232. 1984.

Similar in: Witthoff, T. (ed.). *Five Questions on World Peace*. Stockholm: Women's International League for Peace & Freedom, 117 pp: pp 36–42,117. 1985. (This also published in a Swedish edition by Carlssons, Stockholm.)

#141 Remnants of war. *Ambio* (Stockholm) 13(1):14–17. 1984.

Reprinted in: Goldstein, E.C. (ed.). *Defense. II.* Boca Raton, FL, USA: Social Issues Resource Series, 60 articles: Article 45. 1985.
Summary in: *Scanorama* (Stockholm) 13(5):101. May 1984.
Similar as: Explosive remnants of war: an overview. In: Westing, A.H. (ed.). *Explosive Remnants of War: Mitigating the Environmental Effects*. London: Taylor & Francis, 141 pp: pp 1–16 (Chap. 1). 1985.

#142 Herbicides in war: past and present. In: Westing, A.H. (ed.). *Herbicides in War: the Long-term Ecological and Human Consequences*. London: Taylor & Francis, 210 pp: pp 3–24 (Chap. 1). 1984.

Summary in German in: Dosch, W., & Herrlich, P. (eds). *Ächtung der Giftwaffen*. Frankfurt a.M.: Fischer Taschenbuch Verlag, 192 pp: pp 157–159,191–192. 1985.
Summary in Italian in: Bisanti, L. (ed.). *Gli Erbicidi: Usi Civili e Bellici*. Catania, Italy: Coneditor, 133 pp: pp 65–70. 1985.
Summary in: *Contemporary Issues in Geography & Education* (London) 2(3):42–50. 1987.
In large part in French in: Capdeville, Y., *et al.* (eds). *L'Agent Orange au Viêt-nam: Crime d'Hier, Tragédie d'Aujourd'hui*. Paris: Éditions Tyrésias, 162 pp: pp 75–92. 2005.
Then essentially the same in English in: Bui Thi Hong *et al.* (eds). *Agent Orange in Vietnam: Yesterday's Crime, Today's Tragedy*. Hanoi: National Political Publishing House, 203 pp: pp 92–213. 2008.
Then the original essentially in its entirety in both English and Vietnamese in: Capdeville, Y., *et al.* (eds). 408 pp + 16 pl.: pp 111–132 (in Vietnamese) & pp 296–317 (in English) + pl. 3 & 14. 2009.

#143 *Herbicides in War: the Long-term Ecological and Human Consequences* (Editor & co-author). London: Taylor & Francis, 210 pp. 1984.

#144* Environmental warfare: an overview. In: Westing, A.H. (ed.). *Environmental Warfare: a Technical, Legal and Policy Appraisal*. London: Taylor & Francis, 107 pp: pp 3–12 (Chap. 1). 1984.

Similar in: *Environmental Law* (Portland, OR, USA) 15(4):645–666. Smr 1985.
[***Nb:*** Reproduced in Chapter 6 by permission of the *Stockholm International Peace Research Institute (SIPRI)*, the copyright holder, on 20 March 2012.]

#145 *Environmental Warfare: a Technical, Legal and Policy Appraisal* (Editor & co-author). London: Taylor & Francis, 107 pp. 1984.

Reprinted in Japanese: Tokyo: Hylife Publishing Co., 130 pp. 1986.

#146 The arms race *versus* environmental and human welfare. *Universitat Internacional de la Pau Ponències* (Sant Cugat del Vallès, Spain) 1984:173–177.

Reprinted in: *Man, Evolution, Cosmos* (Sofia) 3(2):50–53. 1984.

#149 Misspent energy: munition expenditures past and future. *Bulletin of Peace Proposals* [now *Security Dialogue*] (Oslo) 16(1):9–10. 1985.

#151 Ban chemical weapons in Europe. *Bulletin of the Atomic Scientists* (Chicago) 41(5):17–19. May 1985.

Reprinted in: *U.S. Dept of Defense Current News* (Washington) 1985(1312):5–12. 20 Jun 1985.
Similar in: *Current Research on Peace & Violence* (Tampere, Finland) 8(1):1–5. 1985.

#154 Towards eliminating the scourge of chemical war: reflections on the occasion of the sixtieth anniversary of the Geneva Protocol. *Bulletin of Peace Proposals* [now *Security Dialogue*] (Oslo) 16(2):117–120. 1985.

#155 The renunciation of chemical and biological weapons. *International Peace Research Newsletter* (Leuven, Belgium) 23(2):15–17. Apr 1985.

#157 *Explosive Remnants of War: Mitigating the Environmental Effects* (Editor & co-author). London: Taylor & Francis, 141 pp. 1985.

#158 The threat of biological warfare. *BioScience* (Washington) 35(10):627–633. Nov 1985.

Preliminary version in German in: Dosch, W., & Herrlich, P. (eds). *Ächtung der Giftwaffen*. Frankfurt a.M.: Fischer Taschenbuch Verlag, 192 pp: pp 73–89,182–192. 1985.

#162 An expanded concept of international security. In: Westing, A.H. (ed.). *Global Resources and International Conflict: Environmental Factors in Strategic Policy and Action*. Oxford: Oxford University Press, 280 pp: pp 183–200 (Chap. 9). 1986.

#163 *Global Resources and International Conflict: Environmental Factors in Strategic Policy and Action* (Editor & co-author). Oxford: Oxford University Press, 280 pp. 1986.

#176* The ecological dimension of nuclear war. *Environmental Conservation* (Cambridge, UK) 14(4):295–306. Wntr 1987.

> Preliminary version in: *Ecoforum for Peace* (Sofia) 1987(1):69–87. [***Nb:*** Reproduced in Chapter 7 by permission of the *Foundation for Environmental Conservation*, the copyright holder, on 26 March 2012.]

#177 Cultural constraints on warfare: micro-organisms as weapons. *Medicine & War* (London) 4(2):85–95. Apr–Jun 1988.

#179 Constraints on military disruption of the biosphere: an overview. In: Westing, A.H. (ed.). *Cultural Norms, War and the Environment*. Oxford: Oxford University Press, 177 pp: pp 1–17 (Chap. 1). 1988.

> Reprinted in: *Transnational Associations* (Washington) 41(3):160–167. May–Jun 1989.

#181 *Cultural Norms, War and the Environment* (Editor & co-author). Oxford: Oxford University Press, 177 pp. 1988.

> Reprinted in part (Chap. 1 & App. 1) in: *Transnational Associations* (Washington) 41(3):160–167. May–Jun 1989.
> Reprinted in Russian as: *[Moral-ethical norms, war, environment]* (Russ. ed., Frolov, I.T.). Moscow: Mir Publishers, 255 pp. 1989.

#182 The military sector *vis-à-vis* the environment. *Journal of Peace Research* (Oslo) 25(3):257–264. Aug 1988.

> Preliminary version in: Goedmakers, A.M.C., *et al.* (eds). *Militaire Aktiviteiten, Natuur en Mileiu. A.* Rijswijk, Netherlands: Raad voor het Milieu- en Natuuronderzoek, 122 pp: pp 5–21. 1987.

#185 Herbicides in warfare: the case of Indochina. In: Bourdeau, P., *et al.* (eds). *Ecotoxicology and Climate: with Special Reference to Hot and Cold Climates*. Chichester, UK: John Wiley, 392 pp: pp 337–357 (Chap. 5.6). 1989.

#186 Chemical and biological warfare: the road not taken. *Peace Research* (Manitoba, Canada) 21(1):17–20,81–83. Jan 1989.

#188 The environmental component of comprehensive security. *Bulletin of Peace Proposals* [now *Security Dialogue*] (Oslo) 20(2):129–134. Jun 1989.

#191 Environmental approaches to regional security. In: Westing, A.H. (ed.). *Comprehensive Security for the Baltic: an Environmental Approach*. London: Sage Publications, 148 pp: pp 1–14 (Chap. 1). 1989.

Reprinted in German in: *Militärpolitik Dokumentation* (Frankfurt a.M.) 14(83–84):5–16,87–90. 1991.

#193 *Comprehensive Security for the Baltic: an Environmental Approach* (Editor & co-author). London: Sage Publications, 148 pp. 1989.

Reprinted in large part in German in: *Militärpolitik Dokumentation* (Frankfurt a.M.) 14(83–84):5–90. 1991.

#195 Proposal for an international treaty for protection against nuclear devastation. *Bulletin of Peace Proposals* [now *Security Dialogue*] (Oslo) 20(4):435–436. Dec 1989.

#196 Comprehensive human security and ecological realities. *Environmental Conservation* (Cambridge, UK) 16(4):295. Wntr 1989.

#202 Towards eliminating war as an instrument of foreign policy. *Bulletin of Peace Proposals* [now *Security Dialogue*] (Oslo) 21(1):29–35. Mar 1990.

#203 Our place in nature: reflections on the global carrying-capacity for humans. In: Polunin, N., & Burnett, J.H. (eds). *Maintenance of the Biosphere*. Edinburgh: Edinburgh University Press, 228 pp: pp 109–120 (Chap. 8). 1990.

#204 Environmental hazards of war in an industrializing world. In: Westing, A.H. (ed.). *Environmental Hazards of War: Releasing Dangerous Forces in an Industrialized World*. London: Sage Publications, 96 pp: pp 1–9 (Chap. 1). 1990.

#206 *Environmental Hazards of War: Releasing Dangerous Forces in an Industrialized World* (Editor & co-author). London: Sage Publications, 96 pp. 1990.

#208 *Effects of Chemical Weapons on Human Health and the Environment*. Nairobi: *United Nations Environment Programme*, Document No. UNEP/GC.16/6 (10 Jan 91), 16 pp.

#209 *Environmental Consequences of Armed Conflict between Iraq and Kuwait* (with *UNEP* staff). Nairobi: *United Nations Environment Programme*, Document No. UNEP/GC.16/4/Add.1 (10 May 91), 10 pp.

#210 We have to seek security in its broadest sense. *World Health Forum* (Geneva) 12(2):137–139. 1991.

(including Arabic, Chinese, French, Russian, & Spanish editions)

#212 Towards a universal renunciation of chemical and biological warfare. *Medicine & War* (London) 7(3):222–223. Jul–Sep 1991.

#213 Environmental security and its relation to Ethiopia and Sudan. *Ambio* (Stockholm) 20(5):168–171. Aug 1991.

#218 *Disarmament, Environment, and Development and their Relevance to the Least Developed Countries* (Editor & co-author). Geneva: United Nations

Institute for Disarmament Research, Research Paper No. 10 (UNIDIR/91/83), xiii + 105 pp. Oct 1991.

#226 Environmental refugees: a growing category of displaced persons. *Environmental Conservation* (Cambridge, UK) 19(3):201–207. Aut 1992.

#232 Protected natural areas and the military. *Environmental Conservation* (Cambridge, UK) 19(4):343–348. Wntr 1992.

#233 Environmental dimensions of maritime security. In: Goldblat, J. (ed.). *Maritime Security: the Building of Confidence*. Geneva: United Nations Institute for Disarmament Research, Document No. UNIDIR/92/89, 159 pp: pp 91–102 (Chap. 6). 1992.

#234 The Environmental Modification Convention: 1977 to the present. In: Burns, R.D. (ed.). *Encyclopedia of Arms Control and Disarmament.* New York: Charles Scribner's Sons, 1692 pp: pp 947–954. 1993.

#236 Biodiversity and the challenge of national borders. *Environmental Conservation* (Cambridge, UK) 20(1):5–6. Spr 1993.

#237 Natural resources, conflict, and security, in a shrinking world. In: Nazim, M., & Polunin, N. (eds). *Environmental Challenges: from Stockholm to Rio and Beyond.* Lahore, Pakistan: Energy & Environment Society of Pakistan -&- Geneva: Foundation for Environmental Conservation, 284 pp: pp 85–116 (Chap. 4). 1993.

#241 The global need for environmental education. *Environment* (Washington) 35(7):4–5,45. Sep 1993.

Reprinted in part in: *Yale Forest School News* [now *Environment Yale*] (New Haven, CT, USA) 80(3):9. Spr 1994.

#243 Human instability and the release of dangerous forces. In: Polunin, N., & Burnett, J. (eds). *Surviving with the Biosphere*. Edinburgh: Edinburgh University Press, 572 pp: pp 307–319 (Chap. 15). 1993.

#244 Environmental security for the Horn of Africa: an overview. In: Polunin, N., & Burnett, J. (eds). *Surviving with the Biosphere*. Edinburgh: Edinburgh University Press, 572 pp: pp 354–357. 1993.

#245 Building confidence with transfrontier reserves: the global potential. In: Westing, A.H. (ed.). *Transfrontier Reserves for Peace and Nature: a Contribution to Human Security*. Nairobi: *United Nations Environment Programme (UNEP)*, 127 pp: pp 1–15 (Chap. 1). 1993.

Shorter version in: *Environmental Awareness* (Vadodara, India) 18(1):11–22. Jan–Mar 1995.

#247 *Transfrontier Reserves for Peace and Nature: a Contribution to Human Security* (Editor & co-author). Nairobi: *United Nations Environment Programme (UNEP)*, xi + 127 pp. 1993.

#257 Population, desertification, and migration. *Environmental Conservation* (Cambridge, UK) 21(2):110–114,109. Smr 1994.

Similar in: Puigdefábregas, J., & Mendizábal, T. (eds). *Desertification and Migrations*. Logroño, Spain: Geoforma Ediciones, 322 pp: pp 41–52. 1995.

#271 Environmental approaches to the avoidance of violent regional conflicts. In: Spillmann, K.R., & Bächler, G. (eds). *Environmental Crisis: Regional Conflict and Ways of Cooperation*. Zürich: Swiss Federal Institute of Technology, Center for Security Studies & Conflict Research -&- Swiss Peace Foundation, Environment & Conflicts Project, Occasional Paper No. 14, 162 pp: pp 148–153. Sep 1995.

Shorter version in: *Environmental Awareness* (Vadodara, India) 20(1):19–24. Jan–Mar 1997.

#276 Dioxins in Vietnam (with Pfeiffer, E.W.). *Science* (Washington) 270(5234):217. 13 Oct 1995.

Reprinted in: *Vietnam Today* (San Francisco) 17(2):1A. Nov 1995.
Reprinted in: *Vietnam Report* (London) 1996(5):4. 1st Qtr 1996.

#277 National security for Eritrea as a function of environmental security. *Environmental Awareness* (Vadodara, India) 18(3):83–87. Jul–Sep 1995.

#284 Unexploded sub-munitions (bomblets) and the environment. *Environmental Awareness* (Vadodara, India) 19(1):5–13. Jan–Mar 1996.

#287 The Eritrean-Yemeni conflict over the Hanish archipelago: toward a resolution favoring peace and nature. *Security Dialogue* (Oslo) 27(2):201–206. Jun 1996.

Preliminary version in: *Expert Meeting on Certain Weapon Systems and on Implementation Mechanisms in International Law*. Geneva: *International Committee of the Red Cross (ICRC)*, 171 pp: pp 75–81. 1994.

#294 Canadian security in a broadened context. *Canadian Defence Quarterly* (Toronto) 26(2):13–14,16,18–19,22. Wntr (Dec) 1996.

#295 Explosive remnants of war in the human environment. *Environmental Conservation* (Cambridge, UK) 23(4):283–285. Dec 1996.

Preliminary version in: Herzig, N. (ed.). *Reverence for Life: Albert Schweitzer and Global Health from Lambaréné to the 21st Century*. Wallingford, CT, USA: Albert Schweitzer Institute for the Humanities, 326 pp: pp 133–139. 1994.

#304 Disruption and manipulation of the environment in times of war. In: IRIS (ed.). *Les Deuxièmes Conférences Stratégiques Annuelles de l'IRIS [Institut de Relations Internationales et Stratégiques]*. Paris: La Documentation Française, 335 pp: pp 187–202. 1997.

#311* Environmental protection from wartime damage: the role of international law. In: Gleditsch, N.P., *et al.* (eds). *Conflict and the Environment*.

Dordrecht, Netherlands: Kluwer Academic Publishers, 598 pp: pp 535–553 (Chap. 32). 1997.

[***Nb:*** Reproduced in Chapter 8 by permission of the *Springer Verlag*, the copyright holder, on 14 March 2012.]

#317 A transfrontier reserve for peace and nature on the Korean peninsula. *International Environmental Affairs* (Hanover, NH, USA) 10(1):8–17. Wntr 1998 (1998–1999).

#318 Establishment and management of transfrontier reserves for conflict prevention and confidence building. *Environmental Conservation* (Cambridge, UK) 25(2):91–94. Jun 1998.

#32. Conflict *versus* cooperation in a regional setting: lessons from Eritrea. In: Suliman, M. (ed.). *Ecology, Politics and Violent Conflict*. London: Zed Books, 298 pp: pp 273–285 (Chap. 13). 1999.

#357* In furtherance of environmental guidelines for armed forces during peace and war. In: Austin, J.E., & Bruch, C.E. (eds). *The Environmental Consequences of War: Legal, Economic, and Scientific Perspectives*. Cambridge, UK: Cambridge University Press, 691 pp: pp 171–181 (Chap. 6). 2000.

[***Nb:*** Reproduced in Chapter 9 by permission of the *Environmental Law Institute*, the copyright holder, on 22 March 2012.]

#365 Environmental values in peace and war. In: Galston, A.W., & Shurr, E.G. (eds). *New Dimensions in Bioethics: Science, Ethics and the Formulation of Public Policy*. Boston: Kluwer Academic Publishers, 225 pp: pp 137–153. 2001.

#371 The environment, its natural resources, and international security: what are the connections?: what are the answers? *Environmental Awareness* (Vadodara, India) 24(2):45–51. Apr–Jun 2001.

#380 Peace, culture and ethics: recent history of conservation values in peace and war. In: Tolba, M.K. (ed.). *Our Fragile World: Challenges and Opportunities for Sustainable Development*. Oxford: EOLSS Publishers -&- Paris: *United Nations Educational, Scientific and cultural Organization (UNESCO)*, 2263 pp (2 vols): pp 865–872 (Vol. 1, Chap. 2.14). 2001.

Preliminary version in: *Environmental Conservation* (Cambridge, UK) 23(3):218–225. Sep 1996.
Summary of preliminary version in: Serageldin, I., & Barrett, R. (eds). *Ethics and Spiritual Values: Promoting Environmentally Sustainable Development*. Washington: International Bank for Reconstruction & Development [World Bank], Environmentally Sustainable Development Proceedings Series No. 12, 54 pp: p. 53. 1996.

#389* Environmental dimension of the Gulf War of 1991. In: Brauch, H.G., *et al.*, (eds). *Security and Environment in the Mediterranean: Conceptualizing Security and Environmental Conflicts*. Berlin: Springer Verlag, 1134 pp: pp 523–534 + 1003–1089 *passim* (Chap. 29). 2003.

[***Nb:*** Reproduced in Chapter 5 by permission of the *Springer Verlag*, the copyright holder, on 14 March 2012.]

#393 Environmental and ecological consequences of war. In: Badran, A., *et al.* (eds). *Encyclopedia of Life Support Systems: Institutional and Infrastructural Resources*. Paris: *United Nations Educational, Scientific and cultural Organization (UNESCO)*, Sect. 1.40.5.1, electronic (www.eolss.net), 20 pp. 2003.

Similar in: Gaan, N. (ed.) *Relevance of Environment: a Critique on International Relation Theories*. Delhi: Kalpaz Publishers, 418 pp: pp 351–366 (Chap. 9). 2005.
Similar in: Geeraerts, G., *et al.* (eds). *Dimensions of Peace and Security: a Reader*. Brussels: P.E.I.-Peter Lang, 283 pp: pp 83–95. 2006.
Similar in: Levy, B.S., & Sidel, V.W. (eds). *War and Public Health*. 2nd edn. New York: Oxford University Press, 486 pp: pp 69–84 (Chap. 5). 2008.

#408 Arthur H. Westing. In: Cruikshank, D.A., *et al.* (eds). *SIPRI at 40: 1966 to 2006*. Stockholm: *Stockholm International Peace Research Institute (SIPRI)*, 156 pp: pp 48–50. 2006.

Part II
Benchmark Papers by the Author: A Selection

Chapter 4
The Second Indochina War of 1961–1975: Its Environmental Impact

Note : *The Second Indochina War of 1961–1975 (known in the USA as the 'Vietnam Conflict') achieved widespread notoriety owing primarily to the damage being inflicted by the USA on the forests and farms of that region (#95, #122)*[1]. *That environmental disruption was achieved primarily via the extensive employment of (a) herbicidal anti-plant chemical warfare agents (#98, #142, #185), (b) bombing and shelling (#63, #71, #149), and (***c***) tractor land clearing (#57, #62); also to a much lesser extent via incendiary attacks (#95, pp 58–60); and finally, quite unsuccessfully via secretly attempted rainfall modification (#95, pp 55–56).*

The history of my own contribution to researching and exposing those anti-environmental attempts to subdue an elusive enemy—attempts subsequently suggested in the language of an international treaty to be employing 'methods or means of warfare which are intended, or may be expected, to cause widespread, long-term and severe damage to the natural environment' (#311, p. 537)—is presented earlier (cf. *Chap. 1*). *The paper reproduced below (#91) serves to summarize that ultimately unsuccessful US strategy and its immediate impact on nature.*[2] *At the time of its publication, the Editor of* Ambio *introduced the paper with the following remarks:*

> *Anti-plant warfare should become an important concern of conservationists, advises the author, whose reports from the Second Indochina War of 1961–1975 in South Viet Nam revealed the extent of the ecological damage done to this area for military purposes. He writes: 'The ecological lessons to be learned from the military tactics employed by the USA in South Viet Nam…are: (a) that the vegetation can be severely damaged or even*

[1] The numbered references are provided in Chap. 3.

[2] Reproduced from: *Ambio* (Stockholm) 4(5-6):216-222; 1975 *with the original title: "Environmental Consequences of the Second Indochina War: A Case Study"*, by permission of the Royal Swedish Academy of Sciences, the copyright holder, on 21 March 2012. Portions of this paper were originally presented at the XIIth International Botanical Congress, Leningrad, July 1975. The author is indebted to Professor Paul W. Richards for useful comments.

A. H. Westing, *Arthur H. Westing*, SpringerBriefs on Pioneers in Science and Practice 1, DOI: 10.1007/978-3-642-31322-6_4,

destroyed with relative ease over extensive areas; (b) that natural, agricultural, and industrial-crop plant communities are all similarly vulnerable; and (c) that the ecological impact of such actions is likely to be of long duration'.

Abstract Limited warfare can result in severe, widespread, and long-term environmental damage. This has been demonstrated by a study of the effects of high-explosive munitions (bombs and shells), chemical anti-plant agents (herbicides), and heavy landclearing tractors ('Rome plows') as employed by the USA in South Viet Nam during the Second Indochina War of 1961–1975 for the purpose of extended large-scale area denial. Although the ecological damage to South Viet Nam was severe, the area-denial techniques used were of doubtful military success. Therefore, should a similar strategy be pursued in some future war, then the ecological damage can be expected to be far worse owing to the military necessity for a greatly expanded application of such techniques.

4.1 Introduction

The means of destruction available to the armed forces of the world are becoming ever more versatile and potent. Today's arsenals contain not only a wide array of anti-personnel and anti-matériel weapons, but also a growing number of anti-plant weapons (herbicides, etc.). Thus, it is not only the enemy soldiers and their fortifications which are subject to ready attack, but also the forest trees providing them with concealment and the agricultural crops providing them with food. Moreover, when the enemy soldiers comprise a guerrilla force—an increasingly common situation in today's world—to strike at their food and cover may seem a particularly attractive military strategy. The consequences of such attack on an enemy through an attack on its vegetation is the primary focus of this paper.

Anti-plant warfare should become an important concern of conservationists inasmuch as it must be carried out over a major fraction of an enemy's territory in order to insure military effectiveness. Such attack could significantly exacerbate the increasingly intolerable strains which, for seemingly unavoidable civil purposes, are already being placed on our earth and its natural resources. It is a problem of particular note since lay people generally do not as yet recognize that methods of anti-plant warfare are now readily available. Moreover, neither the general populace nor civil or military policy makers in positions of responsibility seem as yet to be aware of the serious ecological implications of such attack. Two related humanitarian concerns must be mentioned here as well: (a) no matter what level of adverse effect is achieved against the enemy combatant forces, the enmeshed civil populace is certain to be subjected to a far higher level of privation;

and (b) the various negative effects achieved by such attack will continue to plague the recipient nation long after hostilities cease.

In what follows I outline the nature and severity of the ecological impact that such nominally conventional military techniques as bombing and shelling, chemical herbicide spraying, and tractor clearing have when used as anti-plant weapons. I draw primarily from the experience of the Second Indochina War of 1961–1975—more particularly, from US actions in South Viet Nam—to provide a suggestion of what might be expected in some future counterinsurgency or other local war. It will be seen that these actions have established a dangerous, indeed, frightening, precedent with regard to the systematic devastation of enemy vegetation for military purposes. The following brief descriptions of the South Vietnamese theater of war and of the nature of the war fought there are, however, necessary preludes to this discussion.

4.2 The South Vietnamese Theater of War

South Viet Nam extends over about 17 million hectares of forbidding mountains, gentle hills, and flat plains: it is thus the size of Austria plus Hungary (cf. Figure 4.1). Roughly nine out of ten of its 18 million inhabitants are (or were before the war) peasants, depending for their daily subsistence on what can be gleaned from the land; the vast majority are concentrated in the relatively flat areas. Situated wholly between the Equator and the Tropic of Cancer (ca 19 °N to 17 °N latitude), South Viet Nam is for the most part hot and humid.

The southeast summer monsoons bring high temperatures and a deluge of rain; the northeast winter monsoons are only slightly cooler and less rainy. The geography of the southern portion of the country is dominated by the Mekong River. Its immense delta is covered by rice paddies that, from the air, appear to be vast patchwork quilts (cf. Table 4.1 for the scientific names and families of plants mentioned in the text). The rugged highlands, covering some two-thirds of the country, are characterized by their own patchwork of countless tiny (several hectare) plots. These have been carved out of the jungle for centuries, perhaps millennia, by the primitive hill tribesmen (*Montagnards*) who roam these largely uncharted mountains. Some of the patches support crops; some are too impoverished to support anything but low weeds; however, most are covered by secondary forest growth in various stages of successional development. The more southerly strip of coastal lowlands supports dense mangrove forests (EARI & TVA, 1968; Smith *et al.*, 1967; Williams, 1965).

More specifically, of South Viet Nam's 17 million hectares of land, approximately 57 % is covered by a diversity of upland (inland) forests, 1 % by rubber plantations, 2 % by coastal mangrove forests, 14 % by rice paddies, 3 % by dry-field crops, and the remaining 23 % by a miscellany of types (including grasslands or savannas, reeds, open water, and urban areas). Although the immense array of South Vietnamese higher plant and animal species appears to

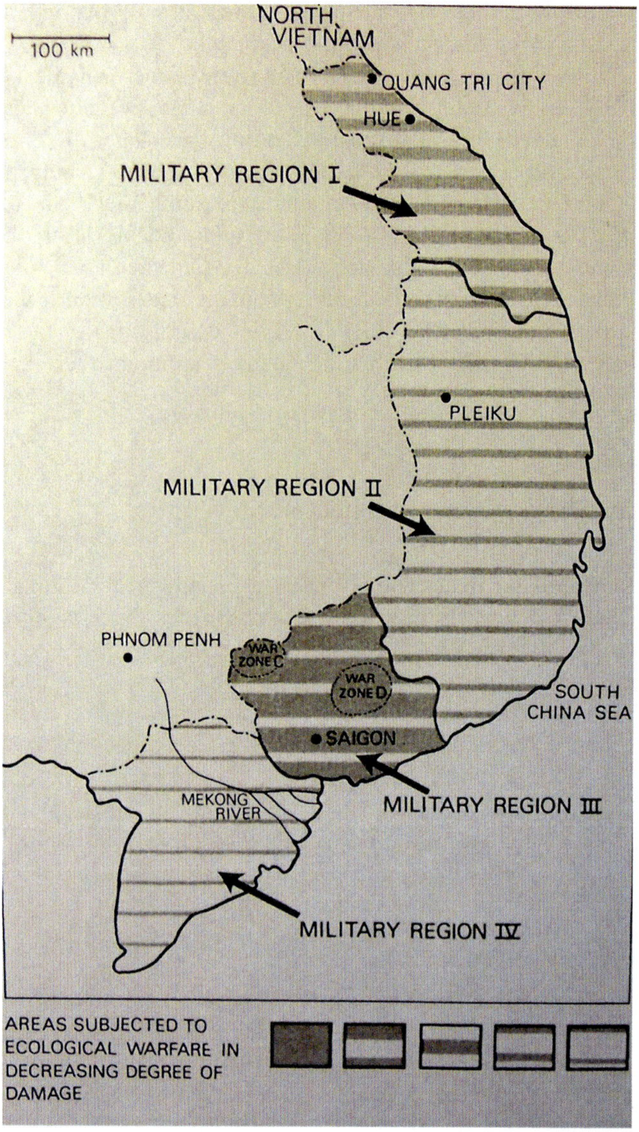

Fig. 4.1 Second Indochina War of 1961–1975: Map of South Viet Nam showing military zones and degree of destruction. © Westing Associates, by permission, AHW 750718

have been quite well catalogued, there exists as yet no adequate ecological analyses and no systematic timber volume inventories.

The several tree-covered areas mentioned above add up to a little over 10 million hectares. Of this combined area, about 56 % can be categorized as dense upland forest ('jungle'), much of it in various states of succession. The dense upland forest

Table 4.1 Scientific Names of Plants Mentioned in the Text

Anisoptera	Dipterocarpaceae	*Melaleuca*	Myrtaceae
Avicennia	Verbenaceae	*Nipa (= Nypa)*	Palmae
Bambusa	Gramineae	*Oxytenanthera*	Gramineae
Cassia	Leguminosae	*Papaya*	Carica papaya,
Coffee	*Coffea arabica*, Rubiaceae	*Pterocarpus*	Caricaceae Leguminosae
Dipterocarpus	Dipterocarpaceae	*Rhizophora*	Rhizophoraceae
Erythrophleum	Leguminosae	Rice	*Oryza sativa*
Guava	*Psidium guajava*,		Gramineae
	Myrtaceae	Rubber	*Hevea brasiliensis*,
Hopea	Dipterocarpaceae		Euphorbiaceae
Imperata	Gramineae	*Sandoricum*	Meliaceae
Jack fruit	*Artocarpus hetero-phyllus*, Moraceae	*Shorea*	Dipterocarpaceae
		Sonneratia	Lythraceae
Kapok	*Ceiba pentandra*, Bombaceae	*Teak*	*Tectona grandis*, Verbenaceae
Lagerstroemia	Lythraceae	Thyrsostachys	Gramineae

type contains a bewildering diversity of dicotyledonous trees, lianas, epiphytes, and herbs as well as some monocotyledons, ferns, and so forth. The tree species vary in height, usually forming two and occasionally three rather indistinct strata (stories); the upper canopy usually attains a height of 20 to 40 meters. The dominant family is the Dipterocarpaceae which is represented by at least 30 major species in the genera *Dipterocarpus*, *Anisoptera*, *Hopea*, and *Shorea*. Another important genus in certain habitats is *Lagerstroemia* in the family Lythraceae. There are also a number of important genera of Leguminosae (*e.g.*, *Erythrophleum*), Guttiferae, and Meliaceae. Moreover, these wildlands support a particularly rich fauna, much of it depending upon an arboreal habitat.

4.3 The US War Strategy in South Viet Nam

The period of major assault upon the vegetation of South Viet Nam dealt with here falls within the eight years between 1965 and 1973. During this extended period, US armed forces attempted to cope with a persistent and highly mobile enemy guerrilla force of perhaps 600,000. Throughout the war the USA maintained physical, on-the-ground control of only a tiny fraction of South Viet Nam; that portion, however, containing in its fragments the various important urban areas of the country and a large majority of its population (Gravel *et al.*, 1971–1972; Huntington, 1967–1968; Hymoff, 1971).

The USA was loathe to commit its army to the sustained ground war (with its attendant high casualties) necessary to achieve a military victory over its enemy. Indeed, its ground force was far too small by traditional standards (by a factor of

Fig. 4.2 Second Indochina War of 1961–1975: Bomb craters in a forest, Bien Hoa Province, South Viet Nam, August 1971. © Westing Associates, by permission, SVN 710808

between three and ten) to attain such an end. The USA attempted to compensate for this deficit and to tilt the military balance in its favor by a variety of means. These included occasional punitive ground raids (the so-called search-and-destroy missions), the employment of technologically sophisticated weapons and techniques, and the lavish expenditure of remotely delivered munitions.

Important in our present context among the several interrelated cost-intensive rather than manpower-intensive means by which the USA attempted to subdue its guerrilla enemy were: (a) forest destruction (primarily to deny the enemy freedom of movement, staging areas, and cover in general); (b) crop destruction (primarily to deny the enemy local sources of food); and (c) a forced relocation of indigenous civilians into the US controlled areas (primarily to deny the enemy local logistical and other support). The employment and consequences of high-explosive munitions (bombs and shells), chemical anti-plant agents (herbicides), and heavy land clearing tractors ('Rome plows') to attain these means of area denial are described below, followed, in conclusion, by the overall implications of such methods of warfare.

4.4 High-Explosive Munitions (Bombs and Shells)

The first category of weapons I discuss consists of high-explosive bombs, shells, and the like (cf. Figure 4.2 & Figure 4.3). Bombs and shells can be used for widespread and long-term area denial so long as they can be procured in sufficiently large quantities and delivered widely and repeatedly. In fact, the Second Indochina War was the first in which this ecologically destructive approach to area

Fig. 4.3 Second Indochina War of 1961–1975: Cratered and sprayed forest, Gia Dinh Province, South Viet Nam, August 1971 © Westing Associates, by permission; SVN 710814

denial was attempted to any significant extent (Kipp, 1967–1968; Littauer & Uphoff, 1972; Westing & Pfeiffer, 1972).

Designating most of the rural portions of South Viet Nam as 'free-fire zones', the USA subjected them to unprecedented amounts of high-explosive munitions. By far the greatest portion of these munitions was expended for purposes of 'harassment and interdiction' and was more or less vaguely targeted. The small amounts of specifically targeted high-explosive munitions expended were divided between direct support of combat missions and the destruction of crops. All told, between 1965 and 1973 the USA dispensed within the borders of South Viet Nam alone some 11 million bombs and a further 217 million or so artillery shells. The majority of the former were the so-called '500- pound' bombs each weighing 241 kilograms, whereas most of the latter were 105-millimeter shells each weighing 13 kilograms. The combined weight of these high-explosive, crater-producing munitions expended by the USA in South Viet Nam finally added up to more than 7000 million kilograms.

Had the high-explosive munitions expended by the USA in South Viet Nam been evenly distributed over that entire country, each hectare would have received some 412 kilograms. However, the vagaries of the war were such that about one-third of the country (Military Regions I plus III, cf. Figure 4.1) was subjected to more than twice this average. And it seems important to stress again that this vast quantity of munitions was directed primarily against the forested and agricultural lands of South Viet Nam; indeed, Quang Tri City was the only important non-rural target ever leveled by the USA in South Viet Nam.

In a forested region the initial detonation of high-explosive munitions is apt to destroy a limited number of wildlife and trees in its immediate vicinity via blast and fragmentation. Beyond this small inner area of quite complete devastation there is a somewhat wider area of partial damage resulting from the flying fragments of metal.

Based on a conservatively calculated average zone of 12.5 m^2/kg of high-explosive munitions in which 50 % of exposed personnel are killed, I am suggesting for reasons indicated below that the zone of environmentally significant damage can be assumed to extend over a similar area. When the above value is applied to the quantity of high-explosive munitions that was, in fact, expended in South Viet Nam, the total area of environmental impact from this source is found to add up to the equivalent of 51 % of the size of the whole country, *i.e.*, to an average of 6.5 % of South Viet Nam per year. The actual geographical extent of this intrusion was, of course, somewhat less than I have just suggested owing to such factors as pattern overlap, topographic irregularities, and forest density. Nevertheless, it is clear that many millions of trees (and lesser numbers of wildlife) living on several millions of hectares were either destroyed outright or struck and wounded by shards of metal.

The puncture wounds in a tree resulting from flying metal provide a ready site of entry for wood-rotting fungi. Particularly in the tropics, the subsequently spreading rot can weaken a tree stem to the point where the wind breaks it off within several years. In some heavily fought-over regions of South Viet Nam (*e.g.*, War Zones C and D, cf. Figure 4.1) an estimated four-fifths of all trees were hit by flying metal. It turned out that species of *Dipterocarpus* and *Anisoptera* as well as the planted rubber are among those quite rapidly vulnerable to destruction of this sort, whereas species of *Hopea* and *Lagerstroemia* are among the more resistant.

The next environmental concern related to the use of high-explosive munitions arises from the craters they leave behind. These holes, which are apt to maintain their topographic integrity for many decades, become a particular ecological concern when they are concentrated in an area. Calculating the dimensions of these craters provides one approach to their environmental impact. Thus the surface area or opening of such a crater averages 0.209 m^2/kg of munitions and the volume of the displaced soil 0.278 m^3/kg. (The average crater of a 241 kilogram bomb—the most frequently used size—has a surface diameter of 8 meters, a surface area of 50 m^2, a maximum depth of 4 meters, and a volume of 67 m^3.)

The craters which were produced in South Viet Nam have a combined surface area of about 148,000 hectares (and thus almost 1 % of the total land surface of the country) and a combined volume (*i.e.*, amount of soil displacement) of 2000 million cubic meters. Particularly in those areas where numerous craters have been blasted out, *i.e.*, in those areas subjected to carpet or saturation bombing, such pock marks on the land can seriously affect the plant community. This occurs especially via the local disruption of the drainage pattern, the disturbance of the water table, and the acceleration of erosional soil loss. For example, each of the 66,000 bombing sorties flown against South Viet Nam by the B-52s alone (the major instrument of carpet bombing) left a crater field averaging 65 hectares in size. The combined area of such disruption just from this source amounted to one-quarter of the land area of the entire country.

Thus, the direct damage from conventional high-explosives to the biota of South Viet Nam, both immediate and delayed, combined with the indirect damage to it via habitat disruption, has resulted in what may well be the most serious (and least recognized) long-term ecological impact of the Second Indochina War.

Fig. 4.4 Second Indochina War of 1961–1975. Sprayed mangrove forest, Gia Dinh Province, South Viet Nam, August 1971. © Westing Associates, by permission, SVN 710815

4.5 Chemical Anti-Plant Agents (Herbicides)

The second category of weapons I discuss consists of the chemical anti-plant agents or herbicides, whose massive military employment was also pioneered during the Second Indochina War (cf. Figure 4.3 & Figure 4.4). Anti-plant agents were employed in Indochina for the denial of forest cover, for the destruction of food plants, and for the decimation of industrial crops (Lang *et al.*, 1974; McConnell, 1969–1970; Westing, 1971–1972).

Great areas of forest and crop land were sprayed from the air in South Viet Nam, particularly during the years 1966 through 1969. The major anti-plant agents for these purposes were either mixtures of the hormone-mimicking compounds 2,4-D [2,4-dichlorophenoxyacetic acid] and 2,4,5-T [2,4,5-trichlorophenoxyacetic acid] or picloram [4-amino-3,5,6-trichloropicolinic acid] (which kill by interfering with the normal metabolism of treated plants) or else the desiccant cacodylic [dimethylarsinic] acid (which kills by preventing treated plants from retaining their moisture content). The most widely used formulation against forests was a mixture of 2,4,5-T plus 2,4-D ['Agent Orange'], applied at the rate of 15 + 14 kg/ha. Another major anti-forest formulation was a mixture of 2,4-D plus picloram ['Agent White'], applied at the rate of 7 + 2 kg/ha. On the other hand, the agent favored for use against crops was cacodylic acid ['Agent Blue'], applied at the rate of 10 kg/ha. Another major anti-crop agent was the 2,4,5-T + 2,4-D formulation already mentioned. All told, about 1.7 million hectares of South Viet Nam were herbicidally treated one or more times, *i.e.,* about one-tenth of the total land area of the country. Some regions were, of course, sprayed more exhaustively than others. For example, in a big rural region largely just north of Saigon (Military Region III, comprising 17 % of South Viet Nam, cf. Figure 4.1) almost one-third of the land was subjected to such attack.

A consideration of the ecological consequences of an herbicidal attack must distinguish not only among vegetational types, but also among numbers of sprayings. I begin with dense upland forest, the most common herbicidal target in South Viet Nam, of which more than 1 million hectares, or 19 %, was sprayed at least once. A single herbicidal attack on dense upland forest results in fairly complete leaf abscission (as well as flower and fruit abscission) within two or three weeks; the surviving trees usually remain bare until the onset of the next rainy season. At that time it becomes evident that there exists a spectrum of sensitivity to the poisons among the many hundreds of tree species which comprise the floristically complex and variable dense upland forest type. From what I could learn, only about 10 % of the trees are killed outright by a single military spraying; the remainder display various levels of injury, as evidenced by differing severities of crown (branch) dieback, temporary sterility, and other symptoms. Among the most sensitive of the dense forest species are *Pterocarpus pedatus* and *Lagerstroemia* spp; among the most resistant are *Cassia siamea* and *Sandoricum indicum*, and among those intermediate between these two extremes are *Hopea odorata*, *Dipterocarpus alatus*, and *Shorea cochinchinensis*.

When the military situation leads to more than one herbicidal attack (as occurred on about one-third of all the sprayed lands), the level of tree mortality increases with each subsequent spraying (more so with briefer intervals between sprayings). Two herbicidal attacks (as occurred on just over one-fifth of the sprayed lands) results in a mortality rate estimated by me to be about 25 %; three such attacks (as occurred on just under one-tenth of the sprayed lands) results in an estimated mortality rate of perhaps 50 %; and four or more such attacks (as occurred on the remaining 4 % of the sprayed lands) results in estimated mortality rates of about 85 % to essentially 100 %.

The first major ecological consideration following an herbicidal attack, particularly in a tropical forest, involves the nutrient-rich leaves which are caused to drop. The newly created leaf litter decomposes rapidly and its nutrients are for the most part lost owing to the dormant or moribund condition of the forest stand (which prevents their recycling) and to the notably poor nutrient-holding capacity of tropical soils. This rapid depletion, which I term 'nutrient dumping', impoverishes the local ecosystem, a condition which takes years or even decades to become rectified by natural processes.

Particularly in those upland areas subjected to three or more herbicidal attacks, some 200,000 hectares of South Viet Nam (1 % of its total land area, 4 % of its total forest lands), a sufficiently high proportion of the extant vegetation is usually destroyed to permit an invasion of the site by a new, relatively impoverished plant community—one of diminished species diversity, biomass, and productivity. This pioneer stage in plant succession is in South Viet Nam often dominated by various relatively low-growing grasses, often by the herbaceous *Imperata cylindrica* or else by a variety of shrubby bamboos in such genera as *Bambusa*, *Oxytenanthera*, or *Thyrsostachys*. These are all tenacious weed species likely to dominate the site for many years or decades into the future (in the case of *Imperata*, particularly if abetted by an occasional fire).

What may not be so readily apparent at first is that herbicidal attacks (particularly if repeated) also raise havoc with the faunal component of the sprayed ecosystems. This is so even if the herbicides used are not in themselves toxic to the animals. Wildlife are simply unable to survive without food or shelter, both of which are largely derived, directly or indirectly, from the plant life of an area.

When a coastal mangrove forest is herbicidally attacked (as occurred on about 150,000 hectares in South Viet Nam, *i.e.*, on about 30 % of all of that country's mangrove type), even a single military spraying most often destroys essentially the entire plant community (cf. Figure 4.4). This as yet inexplicably drastic response applies both to the true mangrove type under daily tidal influence [including such genera as *Sonneratia*, *Avicennia*, *Rhizophora* (a particularly sensitive genus), and *Nipa*] and to the rear (back) mangrove type just inland from the former (including particularly *Melaleuca*). The only partial exception to this taxonomically diverse sensitivity within the mangrove community is *Avicennia*; individuals of this genus growing on the edge of small watercourses do sometimes survive an herbicidal attack. Moreover, and for reasons that are also somewhat elusive, the herbicide-obliterated mangrove sites do not become readily recolonized (owing possibly to inadequate seed source, to destruction of available propagules by crabs, or to other factors). Subsequent soil erosion in destroyed mangrove forests has been found to be severe. The mangrove ecosystem, one of the most productive in the world, also provides the habitat for a rich arboreal fauna. The almost complete loss of this vegetation when attacked with herbicides concomitantly results in the virtually total elimination of the wildlife it supports. Moreover, the mangrove habitat provides breeding grounds or nurseries for most offshore fish and crustaceans, and for many of the freshwater fish and crustaceans as well. Herbicidal mangrove destruction, for this reason, has a debilitating effect on a variety of aquatic fauna, some of commercial importance.

I make only brief mention of the herbicidal attacks on sites supporting agricultural and other crops of economic importance. Although South Viet Nam's paddy (wet) rice lands were largely spared, about 180,000 hectares of its field-crop areas (upland rice, etc.) were subjected to one or more herbicidal assaults, *i.e.*, more than one-third of all such upland areas. Whereas the crops growing on these lands at the time of spraying were obliterated (indeed, the equivalent of more than 300 million kilograms of milled rice), little overt long-range agricultural damage seems to have been done to these sites (particularly if subsequently replanted and fertilized). The herbicides sprayed by the USA were within the span of one growing season either taken out of circulation via their adsorption or binding onto soil particles (the cacodylic acid) or else were decomposed to insignificance (the 2,4-D, 2,4,5-T, and picloram).

The effect of herbicidal attack on rubber plantations was found to vary with tree age and variety (clone, sensu stricto). Whereas all of the varieties used in Indochina are initially defoliated, only the very young individuals are certain to be

Fig. 4.5 Second Indochina War of 1961–1975: Forest removal by Rome plows, Tay Ninh Province, South Viet Nam, August 1971. © Westing Associates, by permission; SVN 710810

killed irrespective of variety. Among the larger plantation trees (those in production), some varieties are almost all killed (*e.g.*, TR 1600, BD 5, and TJIR 1), some seem to recover almost completely (*e.g.*, PB 86), and some are intermediate in their sensitivities (*e.g.*, GT 1, AVROS 50, and PR 107). Similarly among other plants of economic interest, some species proved to be highly sensitive to herbicidal attack (*e.g.*, jack fruit and kapok), some highly resistant (*e.g.*, coffee and teak), and some intermediate (*e.g.*, papaya and guava).

Thus it can be seen that the employment of chemical anti-plant agents or herbicides can readily lead to the serious debilitation of local ecosystems: (**a**) by so-called nutrient dumping; (**b**) by the destruction of the extant vegetational community; and (**c**) by the loss of the animal community, largely via habitat destruction. A decimated plant community on tropical upland sites is likely to become replaced by an ecologically inferior, long-lasting plant community, one with a significantly lesser plant and animal species diversity, a greatly reduced biomass, and a decreased level of productivity. Moreover, a decimated coastal mangrove ecosystem seems to remain desolate for some very lengthy period of time. Finally, when an herbicidal attack is used to destroy either food or industrial crops, this can lead not only to ecological damage, but to social havoc as well.[3]

[3] Not adequately recognized at the time of this writing was that Agent Orange contained small amounts of dioxin, an inadvertent manufacturing impurity, subsequently implicated in serious long-term human (and wildlife) health problems (cf. Chap. 3, #99 & # 276).

4.6 Landclearing Tractors ('Rome Plows')

The final category of weapons I discuss consists of heavy landclearing tractors whose extensive military use was yet another major tactic pioneered during the Second Indochina War (cf. Figure 4.5). The so-called 'Rome plows' were employed for widespread forest removal, crop destruction, and decimation of hamlets and villages. Indeed, perfection of the equipment and techniques involved in this new approach to area denial has been lauded repeatedly by military analysts as one of the most 'striking' and 'exciting' developments of that war (Draper, 1971; Ploger, 1974, pp 95–104; Westing, 1971).

The Rome plow is a 33,000-kilogram armored tractor equipped with a blade designed for the splitting, shearing, and toppling of trees of virtually any size. These mammoth devices were originally brought by the USA to South Viet Nam in 1966 for clearing roadsides (verges) in order to discourage ambushes. By 1968 they were organized into companies of 30 tractors each, whose primary mission was to literally shear off and shove away large forests (often several thousand hectares in size) which were of military advantage to the enemy.

Under routine war-zone operating conditions in South Viet Nam a landclearing company was able to remove heavy jungle (*i.e.*, fully developed dense upland forest) at a sustained rate of 40 hectares per day and light jungle at 160 hectares per day. All told, the USA cleared about 325,000 hectares of South Vietnamese forest land with the Rome plows, *i.e.*, about 3 % of it (or almost 2 % of the total land area of that country). Many thousands of additional hectares of rubber plantations, fruit orchards, and agricultural fields (with their associated irrigation systems) were also obliterated in this fashion. Much of the Rome-plow landclearing was concentrated in a big rural region largely just north of Saigon (Military Region III, cf. Figure 4.1).

The ecological impact of removing virtually all of the vegetation and exposing the soil on thousands of conterminous hectares at a time is a drastic one. The soil immediately becomes subject to massive erosion, particularly in a region of high rainfall and hilly terrain. The soil that does stay in place quickly loses a high proportion of its soluble (available) minerals as a result of the nutrient dumping phenomenon described earlier. Wildlife habitat is destroyed instantly and completely. Moreover, during the period before a new vegetational cover becomes established, the flood-ameliorating capacity of the region is reduced remarkably, a serious deficiency during the heavy rainstorms typical of the summer monsoon season in South Viet Nam. Sooner or later the cleared regions are, of course, repopulated by vegetation. The pioneer plant community that I most often observed to colonize Rome-plowed areas in South Viet Nam was dominated by the pernicious *Imperata cylindrica* or other grasses mentioned previously in connection with herbicide devastation; this in turn can support only a very impoverished animal community.

Thus, it can be seen that the extensive landclearing shown to be feasible with Rome plows leads to locally serious ecological debilitation. The cleared areas undergo severe site degradation and become occupied with long-lasting biotic communities of low plant and animal species diversity, reduced biomass, and diminished productivity.

4.7 The Implications of Anti-Plant Warfare

In the preceding sections I have described briefly several tactical innovations pioneered by the USA in its protracted pursuit of the Second Indochina War. These had the common purpose of subduing a guerrilla enemy by attempting to make vast areas of land continuously inhospitable. This strategy of large-scale and long-term area denial was approached in a number of interrelated ways, none of them, however, involving the taking and holding of this land. Chief among the several alternative—though, in fact, often combined—tactics employed by the USA in South Viet Nam to achieve this end were its attempts: (**a**) to eliminate the cover and concealment that might be provided by forests; (**b**) to eliminate the food and other resources that might be provided by locally grown crops; and (**c**) to eliminate the logistical and other support that might be provided by the indigenous civil population. The various tactics attempted for the attainment of these goals shared not only a common purpose, but also a common effect: that of greatly debilitating or even destroying entire biotic communities for extended periods of time (Falk, 1973; USSR, 1974; Westing, 1974).

In conclusion, among the ecological lessons to be learned from the military tactics employed by the USA in South Viet Nam during the Second Indochina War are: (**a**) that the vegetation can be severely damaged or even destroyed with relative ease over extensive areas—and, of course, with it the ecosystems for which it provides the basis; (**b**) that natural, agricultural, and industrial-crop plant communities are all similarly vulnerable; and (**c**) that the ecological impact of such actions is likely to be of long duration. And finally, the outcome of the war has taught us that the likelihood of military success for such operations is low, indeed; unless, of course, they were to be applied to virtually the entire enemy country—a horrible prospect to contemplate.

References

Draper, S.E. 1971. Land clearing in the Delta, Vietnam. *Military Engineer* (Washington) 63: 257–259.

EARI & TVA [Engineer Agency for Resources Inventories & Tennessee Valley Authority]. 1968. *Atlas of Physical, Economic and Social Resources of the Lower Mekong Basin*. New York: United Nations, 257 pp.

Falk, R.A. 1973. Environmental warfare and ecocide. *Bulletin of Peace Proposals* [now *Security Dialogue*] (Oslo) 4:1–17.

Gravel, M., *et al.* (eds). 1971–1972. *Pentagon Papers: the Defense Department History of United States Decisionmaking on Vietnam.* Boston: Beacon Press, Boston, 632 + 834 + 746 + 687 + 413 pp (5 vols).

Huntington, S.P. 1967–1968. The bases of accommodation. *Foreign Affairs* (New York) 46(4):642–656.

Hymoff, E. 1971. Technology vs guerillas: stalemate in Indo-China. *Bulletin of the Atomic Scientists* (Chicago) 27(9):27-30.

Kipp, R.M. 1967–1968. Counterinsurgency from 30,000 feet: the B-52 in Vietnam. *Air University Review* (Maxwell Air Force Base, AL, USA) 19(2):10–18.

Lang, A., *et al.* 1974. *Effects of Herbicides in South Vietnam. A. Summary and Conclusions.* Washington: US National Academy of Sciences, [398] pp + 8 maps.

Littauer, R., & Uphoff, N. (eds). 1972. *Air War in Indochina.* rev. edn. Boston: Beacon Press, Boston, 289 pp.

McConnell, A.F., Jr. 1969–1970. Mission: Ranch Hand. *Air University Review* (Maxwell Air Force Base, AL, USA) 21(2):89–94.

Ploger, R.R. 1974. *Vietnam Studies: U.S. Army Engineers, 1965–1970.* Washington: US Department of the Army, 240 pp.

Smith, H.H. *et al.* 1967. *Area Handbook for South Vietnam.* Washington: US Department of the Army, Pamphlet No. 550–55, 510 pp.

USSR. 1974. *Prohibition of Action to Influence the Environment and Climate for Military and Other Purposes Incompatible with the Maintenance of International Security, Human Well-being and Health.* New York: United Nations General Assembly, Document No. A/C.1/L.675, 2 + 5 pp.

Westing, A.H. 1971. Leveling the jungle. *Environment* (Washington) 13(9):8–12.

Westing, A.H. 1971–1972. Herbicides in war: current status and future doubt. *Biological Conservation* (Barking, UK) 4(5):322–327.

Westing, A.H. 1974. Proscription of ecocide: arms control and the environment. *Bulletin of the Atomic Scientists* (Chicago) 30(1):24–27.

Westing, A.H., & Pfeiffer, E.W. 1972, Cratering of Indochina. *Scientific American* (New York) 226(5):20–29, 138; (6):7.

Williams, L. 1965. *Vegetation of Southeast Asia: Studies of Forest Types.* Beltsville, MD, USA: US Agricultural Research Service, Publication No. CR 49–65, 302 pp.

Chapter 5
The Gulf War of 1991: Its Environmental Impact

***Note :** The Gulf War of 1991 shares with the Second Indochina War of 1961–1975 the dubious honor of having alerted the world to the enormously destructive impact military actions can have on the environment. Certainly, various wars of the past have also wreaked havoc on the environment* [1]*, and wars of the future will doubtless provide further examples.*

Mostafa K. Tolba, Executive Director of the United Nations Environment Programme was quick to establish an expert task force (of which I was a member and principal author) to investigate the Gulf War, producing under his name one of the earliest authoritative reports on the environmental consequences of that conflict (#209). My subsequent more detailed report on the subject (#389) is reproduced below.[2]

5.1 Introduction

On 2 August 1990, Iraq invaded and annexed neighboring Kuwait, declaring a comprehensive and eternal merger. This action was at once condemned by the *United Nations Security Council* (New York) as a breach of international peace and security (UNSC, 1990a). The invasion of Kuwait by Iraq was also an immediate cause for concern to the *United Nations Environment Programme* (Nairobi), over the resulting destruction of the environment and disruption of social and economic structures (UNEP, 1990).

Repeated subsequent demands and associated actions for immediate and unconditional withdrawal having been to no avail, the *United Nations Security Council* finally authorized Member States to use all necessary means to free

---

[1] The numbered references are provided in Chap. 3.

[2] Reproduced from: Brauch, H.G., *et al.*, (eds). *Security and Environment in the Mediterranean: Conceptualizing Security and Environmental Conflicts*. Berlin: Springer Verlag, pp 523–534 + 1003–1089 *passim* (Chap. 29); 2003 *with the original title "Environmental Dimension of the Gulf War of 1991"* by permission of the Springer Verlag, the copyright holder, on 14 March 2012.

A. H. Westing, *Arthur H. Westing*, SpringerBriefs on Pioneers in Science and Practice 1, DOI: 10.1007/978-3-642-31322-6_5,

Kuwait, failing the withdrawal of Iraq by 15 January 1991 (UNSC, 1990b). In the face of continued noncompliance by Iraq, military actions against Iraq commenced on 17 January, carried out by a coalition of ca 29 Member States (including France, the United Kingdom, and the USA from among the five permanent members of the *United Nations Security Council*). The role of the United Nations in thwarting Iraqi aggression against Member State Kuwait is well described elsewhere (Kaufmann *et al.*, 1991, pp 1–33; Urquhart, 1991), as is its role in the immediate post-war follow-up activities (Ekéus, 1992). It is possible to raise some objections to how the United Nations acted in this matter (Falk, 1991; Schachter, 1991; Weston, 1991).

The present chapter begins with a description of the Gulf War's theater of operations, separately for Iraq, Kuwait, and the Persian (Arabian) Gulf. It goes on to describe the nature of the military assault on the theater of operations, here distinguishing between actions on land (including aerial actions and ground actions) and actions at sea. Discussed next are the environmental consequences of the military actions, both immediate and long-term, separately for air, land, and sea. Brief mention is also made of the societal consequences. The lessons learned from an examination of this war are discussed under two headings, legal and cultural.

This chapter draws upon and extends several relevant earlier studies by the author, including especially one for the *United Nations Environment Programme* that subsequently contributed to a report by its Executive Director (Tolba, 1991), another for the University of Colorado (Boulder, CO, USA) (Westing, 1994), and a third for the *North Atlantic Treaty Organization* (Brussels) (Westing, 1997).

5.2 The Theater of Operations

5.2.1 General

A number of useful publications describing one aspect or another of the Gulf War theater of operations is available (Beaumont, 1978; CIA, 1992, pp 161–163, 188–189; Collins, 1990; Eglin & Rudolph, 1985; Ffrench & Hill, 1971; IUCN, 1985; 1990a; Lindén *et al.*, 1990; Metz, 1990), from which the information presented below in the remainder of Section 5.2 has been gleaned or calculated.

5.2.2 Iraq

Iraq has an area of 434,900 square kilometers; its boundary is ca 3512 km in length, abutting six states and the sea: Iran (1458 km), Jordan (134 km), Kuwait (240 km), Saudi Arabia (686 km), Syria (605 km), Turkey (331 km), and the Gulf (ca 58 km). Iraq's estimated 1990 population was 18.8 million (72 % urban), increasing at the rate of 3.9 % per year (giving a doubling time of 18 years). It committed itself to the Charter of the United Nations on 21 December 1945.

The *Iran–Iraq War of 1980–1988* had a substantial social impact on Iraq and was also environmentally disruptive, especially in rural southeastern Iraq (Karsh, 1989; McKinnon & Vine, 1991, pp 83–89). The inter-war period was of insufficient duration to permit the rural terrestrial habitats to recover from their war-inflicted damage.

Summers in Iraq are hot and dry, winters mild. Dust or sand storms are quite prevalent, especially in summer, and dust fallout onto Iraq approaches the highest in the world. The terrain is mostly broad plains (alluvial plains in the Tigris-Euphrates River system), although there are mountains in the north and northeast and marshes (covering ca 20,000 square kilometers) in the southeast.

Perhaps 1 % of the land of Iraq was in 1990 urbanized or industrialized, 13 % was devoted to agriculture (9 % dry, 4 % irrigated), 9 % was classed as rangeland, 3 % as woodland, and 74 % as desert. Grain production was ca 2.7 million tonnes per year. Iraq, self-sufficient in agricultural production until the 1950s, had become steadily less so since then and in 1990 had to import more than one-third of its food. Large livestock (cattle etc.) numbered ca 2 million; small livestock (sheep etc.) ca 11 million. The waters of the Euphrates and Tigris Rivers (which join to form the Shatt al Arab before flowing into the Gulf) are heavily utilized for irrigation (via an extensive system of dams, dikes, canals, and causeways). Much of the agricultural lands and rangelands suffer from erosion or salinization; much of the water and air are severely polluted. Recoverable resources include oil (production in 1990, ca 160 million cubic meters [ca 1000 million barrels] per year), natural gas (production in 1990, ca 50 million cubic meters per year), phosphates, and sulfur.

At the time of the Gulf War, Iraq had two small (0.5 and 5 megawatt) nuclear research reactors, both at Tuwaitha, ca 25 km south of Baghdad, as well as one or more nuclear-fuel enrichment facilities. It had six huge dams, each at least 15 meters in height and impounding at least 500 million cubic meters of water. There were numerous oil wells, oil storage depots, oil refineries, and petrochemical factories. Press reports in 1990 suggested the existence of two chemical-weapon factories some kilometers northwest of Samarra (in north-central Iraq); the suggestion was additionally made that there was a bacteriological-weapon and toxin-weapon factory at Al-Hakam, several kilometers east of Baghdad. These reports were subsequently confirmed (Feinstein, 1991; Lewis, 2001; Zilinskas, 1997).

Present-day Iraq, which coincides quite closely with ancient Mesopotamia, is particularly rich in **archeological sites** (Zimansky & Stone, 1992a; 1992b). The region has supported a sequence of major cultures and civilizations—Sumer, Akkad, Babylon, and Assyria standing out among them—that goes back some 8000 years or more. Nowhere else in the world is there such a concentration of magnificent relics of early human history. Perhaps the earliest known writings (committed to tablets by Sumerians ca 6000 years ago) have been discovered in

Mesopotamia, quite sophisticated agricultural textbooks among them. The 4500-year-old epic of Gilgamesh was unearthed among the ruins of Nineveh. The legal code of Hammurabi (who may also have built the Tower of Babel) was carved in stone in Babylon perhaps 3700 years ago. Indeed, the region is considered by some scholars to be the very cradle of human civilization. At least several thousand sites within the watersheds of the Tigris and Euphrates Rivers are recognized as being of special value. The 2000-year-old ruins of the Arch of Ctesiphon, ca 30 km southeast of Baghdad, is among the greatest ancient architectural monuments. Iraq has one formally recognized world cultural heritage, the 1800-year-old ruins of the fortifications of Hatra (modern Al-Hadra) southwest of Mosul in northern Iraq. Moreover, the Iraq Museum in Baghdad has housed what was widely considered to be the world's finest collection of ancient Assyrian, Hittite, Phoenician, Sumerian, and Babylonian art; and additionally possessed a renowned collection of Islamic art.

The various rural habitats of Iraq—agricultural, montane, marsh, grassland, arid (desert), etc.—have been more or less intensively utilized for millennia. Population pressures were in 1990 (and remain) such that all of these habitats are to a greater or lesser extent degraded, and are subjected to increasing stress from overutilization and excess pollution. Arid and montane ecosystems are, in any event, notoriously fragile. In 1990 Iraq had no protected natural area registered with the United Nations (IUCN, 1990b, p. 116). Animals native to Iraq and at the time of the Gulf War considered by the *International Union for Conservation of Nature (*IUCN*)* (Gland, Switzerland) to be in danger of extinction included the Persian fallow deer (*Dama mesopatimaca*; IUCN Endangered), Saudi dorcas gazelle (*Gazella dorcas saudiya*; IUCN vulnerable), Dalmatian pelican (*Pelacanus crispus*; *IUCN* Endangered), marbled teal (*Marmaronetta angustirostris*; IUCN Vulnerable), and lesser white-fronted goose (*Anser erythropus*; IUCN Rare) (IUCN, 1990a).

5.2.3 Kuwait

Kuwait has an area of 17,800 square kilometers; its boundary is ca 961 km in length, abutting two states and the sea: Iraq (240 km), Saudi Arabia (22 km), and the Gulf (ca 499 km). Kuwait's estimated 1990 population was 2.1 million (90 % urban), increasing at the rate of 3.8 % per year (giving a doubling time of 18 years). It committed itself to the Charter of the United Nations on 14 May 1963.

Summers in Kuwait are very hot and dry, winters short and cool. Dust or sand storms occur ca 13 % of the time throughout the year, twice as often in summer as in winter, and dust fallout onto Kuwait is perhaps the highest in the world. The terrain is a flat or slightly undulating plain.

Perhaps 1 % of the land of Kuwait was in 1990 urbanized or industrialized, essentially none was devoted to agriculture, 8 % was classed as rangeland, none as woodland, and 91 % as desert. Kuwait must import almost all of its food. A few wadis contain rainwater in the winter (*e.g.*, Wadi al Batin on the western border). Large livestock (cattle etc.) numbered ca 26 thousand; small livestock (sheep etc.) ca 320 thousand. Although there are a few small sources of groundwater, fresh water was (and is) obtained primarily through the large-scale desalinization of seawater. Much of the rangeland suffers from erosion; much of the air is polluted. Recoverable resources include oil (production in 1990, ca 110 million cubic meters [ca 700 million barrels] per year) and natural gas (production in 1990, ca 4800 million cubic meters per year). The Gulf fishery resource was being utilized by Kuwait to some extent.

Kuwait has a considerable number of oil wells (ca 1270) plus a variety of oil storage depots, oil refineries, and petrochemical factories.

The arid habitats that cover most of Kuwait support easily disrupted ecosystems of low productivity. In 1990 Kuwait had no protected natural area registered with the United Nations (*IUCN*, 1990b, p. 124). Animals native to Kuwait and at the time of the Gulf War considered by the International Union for Conservation of Nature to be in danger of extinction included the rhim or Arabian sand gazelle (*Gazella subgutturosa marica*; *IUCN* Endangered), Saudi dorcas gazelle (*Gazella dorcas saudiya*; *IUCN* Vulnerable), Dalmatian pelican (*Pelacanus crispus*; *IUCN* Endangered), and lesser white-fronted goose (*Anser erythropus*; *IUCN* Rare) (*IUCN*, 1990a).

5.2.4 The Persian (Arabian) Gulf

The Persian (Arabian) Gulf—here considered to extend as far as the Strait of Hormuz—has an area of 240,000 square kilometers; its coastline is ca 7530 km in length, touching upon eight states: Bahrain (an island with a coastline of ca 161 km), Iran (ca 2500 km), Iraq (ca 58 km), Kuwait (ca 499 km), Oman (ca 50 km), Qatar (ca 563 km), Saudi Arabia (ca 2510 km), and the United Arab Emirates (ca 1350 km). There are numerous small islands. The Gulf has a mean depth of 35 meters, and (with its area of 240,000 square kilometers) a volume of 8400×10^9 cubic meters. Input is estimated to be 2725×10^9 cubic meters per year, via three routes: inflow through the Strait of Hormuz (2696×10^9 cubic meters per year), stream inflow (5×10^9 cubic meters per year), and rainfall (24×10^9 cubic meters per year). Output, essentially equal to input, is via two routes: outflow through the Strait of Hormuz (2375×10^9 cubic meters per year) and evaporation (350×10^9 cubic meters per year). From these data, replacement time for 50 % of the water can be calculated to be 3.2 years; and replacement time for 90 % of the water, 5.9 years.

Water temperature in the Gulf generally ranges between 15 °C and 35 °C. Mean salinity ranges approximately between 37 and 42 kilograms per cubic meter (the world ocean average being 36 kilograms per cubic meter). There are two tides per day, and the tidal amplitude ranges approximately between 1 and 3 meters. The Gulf has a counter-clockwise current with an average velocity of 20 km per day.

The prevailing winds are north-northwesterly. The Gulf fishery resource was in 1990 harvested to the extent of ca 200 million kilograms per year, a catch that appears to represent over-exploitation. A wide variety of fin-fish are taken as well as shrimp (*Penaeus semisulcatus* etc.) and other shell-fish.

Circa 800 producing oil wells were in 1990 located in the Gulf seabed, which is also crisscrossed with pipelines; there were ca 25 major oil terminals; and oil-tanker traffic was (and remains) heavy (ca 25,000 oil tankers passing through the Strait of Hormuz per year). The routine (and largely intentional) input of oil—by far the worst oil pollution in the world (ca 40-fold higher than any other body of salt water)—is estimated to be ca 180 thousand cubic meters per year. Beaches are contaminated with tar at levels ca 100 times greater than anywhere else in the world. Finally, the *Iran–Iraq War of 1980–1988* had in 1990 still left a substantial legacy of additional environmental damage in the Gulf.

The western side of the Gulf is very shallow, merging into wide tidal mud flats. The intertidal mangrove (*Avicennia*), coral reef, sea-grass (*Halophila* etc.), and various other Gulf habitats all were (and remain) under enormous stress owing to disruptive activities associated, *inter alia*, with oil extraction, tanker traffic, fishery exploitation, and routine oil disposal. Animals native to the Gulf and at the time of the Gulf War considered by the International Union for Conservation of Nature to be in danger of extinction included the green turtle (*Chelonia mydas*; *IUCN* Endangered), hawksbill turtle (*Eretmochelys imbricata*; IUCN Endangered), and dugong or sea-cow (*Dugong dugon*; IUCN Vulnerable); and possibly also the Socotra cormorant (*Phalacrocorax nigrogularis*; IUCN unlisted) (*IUCN*, 1990a).

5.3 Military Assault on the Environment

5.3.1 General

The military actions by the United Nations Coalition forces consisted essentially of: (a) a large build-up in the theater of operations of air, sea, and land forces, beginning in August 1990 (using Saudi Arabia as a major staging area); (b) a massive campaign of aerial bombardment of military targets throughout Iraq and Kuwait, beginning on 17 January 1991; and (c) a full-scale ground assault into Iraq and Kuwait, beginning on 24 February (Feinstein, 1991; Fotion, 1991; Joffe, 2000; Lopez, 1991; Luttwak, 1994; Trux, 1991; Walker & Stambler, 1991). As a result, Iraqi forces were essentially driven out of Kuwait by 26 February, and all combat operations ceased on 28 February.

Both the aerial and ground actions are described next, but first it should be noted here that no unconventional weapons (chemical, toxin, biological, or nuclear) were employed by any of the belligerents during this war, although this had been a concern because the armed forces on both sides were assumed to have all or some of those capabilities. Indeed, some of the armed forces comprising the

United Nations Coalition were vaccinated against the botulinal toxin, a generally lethal antigenic protein produced by the bacterium *Clostridium botulinum*; and antibiotics were stockpiled by those forces in the event of an attack with the live spores of *Bacillus anthracis*, the causative agent of anthrax, generally lethal in its pulmonary form.

5.3.2 On the Terrestrial Environment

5.3.2.1 Via Aerial Actions

The United Nations Coalition forces carried out one of the most intense campaigns of aerial bombardment in military history, directed against Iraqi military targets throughout Iraq and Kuwait, in urban, industrial, and rural areas—with more than 2000 individual aircraft ('sorties') per day going out on bombing runs, more than 40,000 in total. Some of those aircraft were fighter-bombers with a capacity of only several bombs, generally directed against point targets. Others were heavy bombers (the 'B-52 stratofortresses') with a much larger bomb capacity, generally directed against area targets. As then fitted out, each of these bombers had a capacity of ca 12,000 kilograms (*e.g.*, 50 high-explosive fragmentation bombs of ca 240 kilograms a piece). Most of the bombs expended were conventionally aimed, thus actually striking their intended target perhaps 20 % of the time; the very small remaining fraction were so-called 'smart' bombs, which strike their intended target perhaps 60 % of the time.

Many of the high-explosive *fragmentation bombs* being used were one of three traditional types and sizes: '500 pound' bombs (actual weight, ca 240 kilograms; explosive content, ca 90 kilograms); '750 pound' bombs (actual weight, ca 360 kilograms; explosive content, ca 180 kilograms); and '2000 pound' bombs (actual weight, ca 890 kilograms; explosive content, ca 430 kilograms). The extent of environmental disruption of a bomb depends not only upon its size, but also upon such factors as type of bomb fuze, character of the attacked terrain, and nature of the impacted flora and fauna. A rough over-all estimate of the extent of complete site destruction is 12.5 square meters per kilogram of actual bomb weight. Thus, for a '500 pound' bomb, the area of destruction would be ca 0.30 hectare. Moreover, a typical strike zone for a 50-bomb drop from a 'B-52' bomber would cover ca 30 hectares; since these bombers usually flew in sets of three, the combined area would come to ca 90 hectares.

Also being used to some extent were fuel/air explosive *concussion bombs*. These bombs are filled with ethylene oxide or other readily volatile and combustible liquid. The canister is caused to burst near the ground, the released fuel being allowed to mix briefly with the air before detonation. The resulting blast wave (shock wave) produces a transient over-pressure of ca 2000 kilopascals (ca 20 atmospheres). This produces no crater, but does destroy any plants or animals within its zone of effectiveness—for a 1000 kilogram bomb, an area of ca

1 hectare. It should be added that malfunctioning fuel/air explosive bombs might produce a fireball rather than a blast wave. Also occasionally used was a huge second type of concussion bomb, the explosive charge of which is a gelled aqueous slurry of ammonium nitrate and aluminum powder. The environmental impact of this blast weapon is comparable to that of the fuel/air explosive bomb.

Bombing has, of course, become a routine form of combat during the past half century. Nonetheless, it must be noted that the 82,000 tonnes or more of high-explosive bombs expended by the United Nations Coalition forces during this brief war (98 % of which was by the USA) represented a level of intensity (in spatial plus temporal terms) that surpassed that of the Allied bombing campaign during World War II, of the United Nations Command (primarily US) bombing campaign during the Korean War of 1950–1953, and of the US bombing campaign during the Second Indochina War of 1961–1975 (Westing, 1985b). To recapitulate, in the Gulf War, the United Nations Coalition forces carried out one of the most successful campaigns of aerial bombardment up to that time in military history, directed against Iraqi military targets throughout Iraq and Kuwait, in urban, industrial, and rural areas.

Aside from the destruction of military targets by the bombing campaign (command centers, weapons and their delivery systems, fortifications, military transportation and communication facilities, armed forces, etc.), there was also a considerable level of unavoidable collateral (incidental) damage, both immediate and delayed. Included among the collateral casualties of bombing were not only civilians, but also municipal, industrial, agricultural, and other accouterments of civilian life, cultural heritages, and the natural environment. Moreover, some fraction of the bombing would seem to be difficult to justify in terms of tactical relevance (*i.e.*, of military necessity).

In addition to any direct damage from bombs, the possibility exists of a release of so-called *dangerous forces* following an attack on certain targets, whether intended or not (Westing, 1990a). The dangerous forces that have become ever more likely to be released over wide areas in wartime now include radioactive gases or aerosols from nuclear facilities, toxic gases or aerosols from chemical facilities, impounded waters from hydrological facilities, and, under some conditions, pathogenic micro-organism from microbiological facilities or urban sewerage systems. Indeed, some of the air attacks were directed against both nuclear and chemical facilities in Iraq, although apparently with only minimal levels of potentially damage-amplifying releases. Among the targets singled out for destruction by bombing and missile attack were the two operating nuclear reactors at Tuwaitha (Feinstein, 1991).

Iraqi forces emplaced over 500 thousand *land mines* (anti-personnel, anti-vehicle) in rural Kuwait and over 1 thousand *sea mines* in its adjacent waters. The *United Nations Security Council* demanded of Iraq to provide all information in identifying the locations of these explosive remnants to facilitate their clearance (*UNSC*, 1991a, ¶3.d). United Nations Coalition forces expended huge amounts of high-explosive munitions, including *cluster bombs*, a substantial fraction of which did not explode as intended at the time of use, adding significantly to the post-battle residues of unexploded ordnance.

Also to be mentioned is the expenditure, primarily by the USA (and to a minor extent also by the United Kingdom), of *armor-piercing munitions* rendered more effective through the incorporation of the highly dense metal uranium after having been depleted of much of its radioactivity (in essence, depleted of about two-thirds of the uranium-235 isotope so that what is used consists mainly of the uranium-238 isotope). Data released by the USA suggest that it fired such munitions from aircraft containing a total of more than 250,000 kilograms of depleted uranium, and containing at least a further 50,000 kilograms fired from tanks and other ground-based weapons.

Of grave concern in this theater of operations was the release—onto the land or into the sea—of massive quantities of *oil* from attacked wells, pipelines, storage structures, refineries, and ships (tankers etc.); and also the release into the atmosphere of large amounts of soot from oil fires. The destruction of oil facilities is also nothing new to warfare (Westing, 1980, pp 165–167). During World War I, Allied (British) forces destroyed Romanian oil fields, doing so in order to hamper the Axis (German) war effort (Yergin, 1991, pp 179–182). Then during World War II, fully 15 % of all Allied bombing in the European theater of operations was directed against the oil facilities to which Germany had access, again to deny the oil to Germany (Heinebäck, 1974, pp 142–145). And ca 25 % of all ships that the USA lost during World War II were oil tankers (Westing, 1980, pp 165–166). During the *Iran–Iraq War of 1980–1988*, both sides destroyed numerous enemy oil facilities, in these cases apparently largely for punitive purposes, which resulted in several major pollution episodes (McKinnon & Vine, 1991, pp 83–86). Indeed, during the Iran–Iraq War the *United Nations Security Council* called upon both states parties to refrain from any action that could endanger marine life in the region of the Gulf (UNSC, 1983, ¶5). However, the Iraqi attacks against Kuwaiti oil facilities during the Gulf War seemed not only to be in essence tactically irrelevant (*i.e.*, not justified by military necessity), but were also unprecedented in both their scope and impact.

5.3.2.2 Via Ground Actions

The Iraqi *oil releases* into the environment resulted primarily from the sabotaging of ca 730 oil wells (some reports suggest even more), of ca 20 collecting centers, and of 3 or more oil tankers. Many of the sabotaged oil wells continued to discharge oil for months, the last ones not being brought under control until November 1991. The huge resulting releases into the environment—perhaps of the order of 10 million cubic meters (of the order of 60 million barrels), the published estimates varying wildly—took two major forms: (a) as a *liquid* from damaged oil wells, pipelines, storage tanks, and oil tankers, impinging on both terrestrial and marine environments; and (b) as *smoke* from oil fires from ca 650 oil wells, that is, as soot and various combustion gases, impinging upon the atmosphere and, as fallout, on terrestrial and marine environments (Bakan *et al.*, 1991; Browning *et al.*, 1991; Hobbs & Radke, 1992; Johnson *et al.*, 1991; Limaye *et al.*, 1991; Small, 1991).

Huge numbers of *off-road vehicles* are used by ground forces deployed in the field, both in preparation for and during land (ground) warfare. These items of military equipment include tanks (more than 1700 US tanks alone during the Gulf War), self-propelled artillery, armored personnel carriers, trucks, tractors, and a variety of lighter vehicles. In traversing the countryside, they destroy the vegetation and disrupt the soil surface, with highly detrimental effects on the local ecosystems. Desert habitats are especially vulnerable to such disturbance. Thus, the habitat disturbance resulting from the Battle of El Alamein of October–November 1942 increased the number of dust storms in northern Egypt by ca 10-fold and also their severity (Oliver, 1945–1946). The problem there persisted for the several subsequently undisturbed years that it took for the re-establishment of a soil-stabilizing vegetative cover. More subtle effects on the desert habitat of battle disruption persist for decades.

5.3.3 On the Marine Environment

The marine environment is by no means immune from military disruption (Westing, 1980, pp 152–169). Indeed, during the Gulf War, huge amounts of oil were released into the Gulf, apparently of the order of 1 million cubic meters (of the order of 6 million barrels). Some of this was an unavoidable and incidental result of a war waged in a theater of operations containing huge numbers of operating oil wells, both on land and offshore, plus all of the associated pipelines, collecting centers, refineries, and tankers. However, most of the oil that was discharged into the waters of the Gulf was done so intentionally by the Iraqi forces.

5.4 Environmental Consequences of the Gulf War

5.4.1 General

The environmental consequences of the Gulf War became a matter of immediate concern to the United Nations (*e.g.*, Karrar *et al.*, 1991; Tolba, 1991; UNEP, 1991; 1991–1992; *UNGA*, 1991a; 1991b; 1992a; *UNSC*, 1991b, ¶16) as well as to various governments (*e.g.*, Gulf Task Force, n.d.; Lee, 1992), nongovernmental organizations (*e.g.*, Greenpeace, 1992), and individuals (*e.g.*, Hawley, 1992; McKinnon & Vine, 1991). There was also an early outpouring of publications on the subject in the popular and semi-popular literature (*e.g.*, Barnaby, 1991; Canby, 1991; Cloudsley-Thompson, 1991; Earle, 1992; Oza, 1991; Pope, 1991; Price,

1991; Renner, 1991; Seager, 1992; Sheppard & Price, 1991; Warner, 1991; White, 1991; Wolkomir & Wolkomir, 1992). However, some of those early publications were more or less ill informed, hastily or prematurely prepared, or polemical in nature. A number of more recent analyses of varied quality are now also available (*e.g.*, Abdulraheem, 2000; Al-Awadi, 1995; Brown & Porembski, 2000; Charrier, 1998; Hegazy, 1997; Horváth & Zell, 1996; Joffe, 2000; Khuraibet, 1999; Omar *et al.*, 2000).

The social and environmental consequences of the combat operations of the Gulf War were truly formidable. As to the environmental impact, two forms of devastation stand out: (a) damage resulting from the *massive bombing* campaign mounted by the United Nations Coalition forces; and (b) damage resulting from the *massive destruction of oil facilities* by the Iraqi forces. Outlined next are the impacts of these assaults on the atmosphere, the terrestrial environment, and the marine environment; also touched upon are the impacts on society.

5.4.2 On the Atmosphere

The many hundreds of oil well fires set by Iraqi forces in Kuwait led to the generation of huge amounts of smoke (soot and various combustion gases). The continuously replenished smoke pall persisted in the area for several months, during that period reducing the amount of incoming sunlight and lowering ambient temperatures (cf. Figure 5.1).

The soot fallout deposited in the region by the smoke combined with some oil aerosols (together, so-called 'black rain') killed or damaged some vegetation. Adverse health effects from the smoke were also reported, especially for those already suffering from respiratory ailments or who were otherwise in a frail condition (cf. below); it presumably had a similar effect on the regional wildlife, both on land and in the Gulf. Finally, the burning oil contributed to the excess carbon dioxide in the atmosphere, that so-called 'greenhouse' gas thereby exacerbating the global-warming problem.

5.4.3 On the Terrestrial Environment

The desert environment (*i.e.*, one too dry to support either a woodland or grassland ecosystem) characterizes much of the rural areas in the war zone. The ecosystems that develop under arid conditions have an easily degraded soil that supports a sparse and highly specialized biota. Such a system is readily damaged and very slow to recover. Where water is available in an arid region (from springs or exotic streams) a richer biota is present, but such sites are rare and for the most part heavily utilized for crops or livestock, both also readily subject to severe

Figure 5.1 Gulf War of 1991: Smoke from oil well fires, Kuwait, April 1991. © US Department of Defense, by permission: KWT 910422.

disruption. Irrigation systems (which require considerable routine maintenance under any conditions) are very vulnerable to damage.

The major long-term effects of the Gulf War on the terrestrial environment have been depletion of the sparse vegetation, soil compaction, disruption of the soil surface (leading to the accelerated development and movement of sand dunes, and soil contamination (in some cases apparently down as far as the groundwater).

The many hundreds of oil wells and associated facilities sabotaged by Iraqi forces in Kuwait resulted in the release of large amounts of oil into the desert environment. Some 200 or more small lakes of oil were created (of which only a small number were subsequently fully drained), leading to various environmental problems, presumably including the potential at least for percolation downward to groundwater. Somewhat more than 25 % of the Kuwaiti desert is estimated to have been befouled to varying levels by oil (Brown & Porembski, 2000; Omar *et al.*, 2000). Areas of tar (hardened oil) and their surroundings have been gradually revegetating as they began to fragment, disintegrate, and become covered with sand (Brown & Porembski, 2000; Hegazy, 1997; Omar *et al.*, 2000). The break-up of the tar-covered areas occurs largely as a result of off-road vehicular traffic, livestock trampling, and burrowing animals (ants, lizards, rodents). On the other hand, once revegetation begins, for it to be successful livestock would have to be excluded from the area (something being done only rarely).

One of the sad effects of the many remaining oil lakes is that birds and insects mistake them for pools of water, to become mired and killed in the oil (Horváth & Zell, 1996; Omar *et al.*, 2000). The oil on the surface generally penetrates downward for only several centimeters, but in some cases is said to go down more deeply, thereby perhaps contaminating groundwater. Recovery of the desert wildlife to its prewar status will depend largely on recovery of the desert flora. Some reintroductions are being considered, especially in the few protected natural areas.

5.4.4 On the Marine Environment

The [Persian Gulf] Regional Organization for the Protection of the Marine Environment [ROPME] (Safat, Kuwait), with the assistance of the International Maritime Organization (London), coordinated some immediate postwar oil recovery efforts from Gulf waters as well as some littoral clean-up, carried out by several states which had volunteered such services (Al-Awadi, 1995). However, a subsequent longer term rehabilitation plan prepared in 1991 by the *United Nations Environment Programme* at the request of the Regional Organization did not materialize at the time for lack of requested funds from the eight littoral or other states (Al-Awadi, 1995). In 1992, in partial response to the *United Nations Environment Programme* plan, the US National Oceanographic and Atmospheric Administration did carry our a three-month ship-based oil-damage assessment of the Gulf with the assistance of an international team of scientists, with remote-sensing support from a US National Aeronautics and Space Administration orbiting satellite. Three brief further surveys were conducted during 1993-1994 by a Japanese research vessel.

Of the oil discharged into the Gulf during the Gulf War, ca 40 % evaporated, ca 10 % dissolved, and the remainder floated (Abdulraheem, 2000). Of the floating oil, ca 20 % was recovered, ca 50 % washed ashore, and much of the remainder sank to the bottom. Thus, oil slicks severely contaminated some offshore waters plus ca 400 km of coastline, primarily Saudi Arabian (Readman *et al.*, 1992), disrupting marine habitats, at the time, *inter alia*, killing much migratory marine wildlife (primarily avian, mammalian, and reptilian). The soot fallout onto the Gulf from the oil fires contributed to the death and disruption of the marine biota. The postwar shrimp population in the Gulf declined by ca 90 % from prewar levels (Abdulraheem, 2000), but the annual shrimp harvest by Kuwait regained its approximate prewar levels within less than seven years (Charrier, 1998). However, long-term recovery of the Gulf ecosystems and the biota they support is dependent not so much on the wartime insults, which have now largely merged indistinguishably with the chronic ambient pollution load (which, as noted earlier, is the worst in the world), but rather on a general environmental control and restoration program.

5.4.5 *On Society*

Although not the focus of this study, it must be noted that the impacts on both the Kuwaiti and Iraqi civil sectors were massive. As to **Kuwait**, the Iraqi invasion and occupation of that state led to enormous levels of death and injury, destruction, pillaging, generation of refugees, and related social chaos (Ahtisaari *et al.,* 1991b; Al-Awadi, 1995; Ascherio *et al.*, 1992; Geiger, 1994; Michel, 1991). One deplorable outcome of the occupation of Kuwait, of peripheral environmental concern, was the terrible cruelty inflicted by Iraqi forces upon the animals in the Kuwait zoo (McKinnon & Vine, 1991, pp 94–96; Pawlick, 1991). As to **Iraq**, the bombing campaign mounted by the United Nations Coalition forces against that state was, as already noted, extraordinarily intensive, with the collateral impact on the Iraqi civilian population and its urban and industrial artifacts thus immense (Ahtisaari *et al.,* 1991a; Lee & Haines, 1991; Michel, 1991). The use by the United Nations Coalition forces (primarily by the USA) of munitions containing depleted uranium has raised a public-health concern owing to the residual contamination of the battlefield of this highly toxic and radioactive heavy metal (Birchard, 1998), so that the *World Health Organization* (Geneva) has initiated an epidemiological survey to investigate that possibility (Abbott, 2001; Kapp, 2001).

The *Kuwaiti infrastructure*, both urban and industrial, was essentially rebuilt within two years of the end of the war (Charrier, 1998). For example, by 1993 Kuwaiti oil production was ca 5 % above its prewar level, *i.e.*, ca 115 million cubic meters (ca 730 million barrels) per year. This rate of recovery was comparable to that of western Europe following World War II (Westing, 1980, pp 61–63). The very modest agricultural sector of Kuwait was said to have been devastated by Iraqi military actions, especially at Abdali and Wafra (Omar *et al.*, 2000). As noted above, annual shrimp harvest from the Gulf by the Kuwaiti fishing fleet regained its approximate prewar levels within less than seven years. Now Kuwait is about to embark on a rural rehabilitation (bioremediation) effort beyond what was needed to restore oil production, expected to cost of the order of US$ 1 thousand million (Shouse, 2001). It will be of interest to note that, all told, Kuwait has filed a series of damage claims with the United Nations Compensation Commission (Geneva) that total ca US$ 300 thousand million (of which ca US$ 40 thousand million is for environmental claims). Neighboring states have filed additional damage claims adding up to ca US$ 45 thousand million.

Most mines and other *unexploded ordnance* in the urban and industrial areas of Kuwait were cleared by 1993 (Charrier, 1998; Khuraibet, 1999; Omar *et al.*, 2000). Mines and other explosive remnants of the Gulf War in Kuwait led to hundreds of human casualties during the first five postwar years (Khuraibet, 1999). However, substantial progress in the clearance of mines and other unexploded ordnance was achieved throughout the country by 1995—perhaps as much as 85 % complete (Omar *et al.*, 2000)—so that human explosive-remnant casualties are now quite low. Indeed, the Kuwaiti Ministry of Defence reported having neutralized over 1.6 million mines (and presumably other unexploded ordnance) by the end of 1996 (Omar *et al.*, 2000). On the other hand, the remaining mines and other explosive remnants are sure to continue to cause deaths and injuries among humans and livestock for many years into the future (Westing, 1985a; 1996).

The precious *archeological heritage* of Iraq, both that portion formerly housed in the Iraq Museum in Baghdad and that portion *in situ* at numerous field sites was providentially largely spared wartime military destruction. Tragically, however, several thousand precious artifacts in the Museum had been removed for safe-keeping early on, but were then looted and lost to society during a period of immediate postwar chaos, as were also various important field sites (Lawler, 2001).

5.5 Some Lessons Learned from This War

5.5.1 General

Environmental disruption is an inevitable and occasionally dramatic concomitant of war (Westing, 1980). In many instances, this combat-associated disruption of the environment is an *incidental* outcome of hostile military actions, for example, in the Second Sino–Japanese War of 1937–1945 (Westing, 1977, pp 54, 62–63). In other instances, it is an *intentional* component of the strategy of a belligerent, for example, in the Second Indochina War of 1961–1975 (Westing, 1976; 1989). On the other hand, societal attitudes or cultural norms do exist that tend to limit such disruption, whether it be incidental or intentional. Moreover, those cultural norms have to some considerable extent been translated into legal norms, formal constraints that have thus found their way into one realm or another of international law. Of particular relevance here is the Law of War (Law of Armed Conflict; International Humanitarian Law); and also, although still of lesser direct relevance, International Environmental Law. The disparate relevant legal restraints from these realms are briefly examined below, with the Gulf War in mind, followed by an examination of the cultural norms that underpin them. Indeed, the Gulf War has made it easy to offer some suggestions for minimizing environmental disruption in time of war. These suggestions can be readily divided into those concerning legal norms and those concerning cultural norms. But in creating such a dichotomy it must be stressed at the outset that legal norms would not exist, and could not function, in the absence of supportive cultural norms. Legal norms, in turn, serve to reinforce, and help to spread, their underlying cultural norms. Thus, each of these two regimes supports the other.

5.5.2 Regarding the Legal Regime

Substantial numbers of authorities have analyzed international law as it relates to the environmental component of the Gulf War (*e.g.*, Arkin *et al.*, 1991, pp 114–144; Boutros-Ghali, 1992; Bouvier, 1991; Casey-Vine, 1991; Fauteux, 1991; 1992a; 1992b; Goldblat, 1991; Green, 1991; Lee, 1992, pp 47–74; Plant, 1992; Roberts, 1993; Robinson, 1991; Sandoz, 1992; Terry, 1992; York, 1991; Westing, 1997; Zedalis, 1991).

There is no question that, during the Gulf War, the USA and other United Nations Coalition Forces over the course of some weeks caused environmental damage through their extraordinarily intensive bombing campaign, but it has been made clear above that it was the massive releases by Iraq of Kuwaiti oil that made this war notorious from an environmental standpoint. In short, the release of oil in liquid form befouled large areas of Kuwait's terrestrial environment as well as inshore portions of the Gulf. Further huge amounts of escaping oil were set on fire, thereby saturating the local atmosphere with dense, noxious smoke for a period of months. The protection of the natural environment deriving from 1977 Bern Protocol I (*UNTS* 17512) would have been relevant, but although Kuwait was a state party during the Gulf War, neither Iraq nor the USA were. What protection the 1980 Land Mine Protocol (*UNTS* 23495) might have afforded was also not applicable since none of the three relevant states were at the time states parties to that treaty. More generalized (non-specific) restraints of relevance are touched upon next.

The environmental destruction in the Gulf War was presumably covered, by virtue of being 'enemy property', by 1899 Hague Convention II and 1907 Hague Convention IV as well as by 1949 Geneva Convention IV (*UNTS* 970), which, *inter alia*, proscribe the destruction of enemy property during hostilities, unless it be imperatively demanded by the necessities of war (both Hagues, Annex Article 23.g), or during occupation (both Hagues, Annex Article 55; Geneva, Article 53). Of the three main states involved in the Gulf War (Iraq, Kuwait, and the USA), only the USA was a state party to the two Hague Conventions, but it is generally accepted that this pair of treaties has achieved a status of 'customary' international law, that is, to be unavoidably binding upon all states. Geneva Convention IV is also applicable since all of the three main states involved were states parties during the Gulf War (and, in any case, it is also generally considered to have achieved 'customary' status). It is thus important to note that the *United Nations Security Council* resolved that Iraq was liable for any direct environmental damage and depletion of natural resources caused by its invasion and occupation of Kuwait (*UNSC*, 1991b, ¶16), presumably on the basis of 1907 Hague Convention IV (Article 3) (cf. also 1977 Bern Protocol I, Article 91). The Security Council further demanded of Iraq (apparently without specific legal basis) that it supply information on the locations of the land mines and similar devices it had emplaced in Kuwait, so as to facilitate their clearance (*UNSC*, 1991a, ¶3.d). As an aside, it is of interest to note here once again that what may have been the first formal recognition by the Security Council of wartime environmental damage occurred during the Iran–Iraq War of 1980–1988, in which it called upon both states parties to refrain from any action that could endanger marine life in the region of the Gulf (*UNSC*, 1983, ¶5).

5.5.3 Regarding the Cultural Regime

The adoption, interpretation, and observance of the legal norms just summarized depend upon the societal attitudes that they reflect, both as to overall attitudes regarding acceptable forms of interstate conflict resolution and as to specific attitudes regarding acceptable forms of environmental exploitation and disruption (Westing, 1988). The continued acceptance of warfare as a means of conflict resolution, and the widespread refusal to submit to the compulsory and unconditional jurisdiction of the International Court of Justice (The Hague), are the basic causes of wartime damage to the environment, but are beyond the scope of this discussion (Westing, 1990b). On the other hand, the increasingly widespread and ever heightening concern over the state of the human environment is of central importance here.

The just described existing legal norms protective of the environment in time of war, even if accepted by a belligerent, are open to a considerable range of interpretation. They, in turn, derive from the laudable—though even more nebulous—fundamental concept of the Law of War that *the right of belligerents to choose methods of warfare is not unlimited.*

The legal norms also gain some further strength from the emerging basic concept of International Environmental Law that *states do not have the right to cause environmental damage beyond their own borders.* Perhaps the most succinct enunciation of this principle of respect can be found in the non-binding 1972 United Nations Declaration on the Human Environment, which proclaims that 'States have…the responsibility to ensure that activities within their jurisdiction or control do not cause damage to the environment of other States or of areas beyond the limits of national jurisdiction' (UNGA, 1972b, Principle 21)—subsequently singled out for endorsement by the *United Nations General Assembly* (*UNGA*, 1972a). The notion gains some strength from the 1963 Partial Test Ban Treaty (*UNTS* 6964), which prohibits certain forms of transboundary radioactive air pollution (Article 1); and also from 1907 Hague Convention V, which establishes that in the case of land war the territory of neutral states is inviolable (Article 1). Further modest support derives from the 1972 World Heritage Convention (UNTS 15511), via which it is agreed not to take any deliberate measures which might damage, directly or indirectly, any cultural or natural heritage situated on the territory of other states parties (Article 6.3). The most recent support has been provided by the 1992 Biological Diversity Convention (UNTS 30619), via which the many states parties have formally accepted 'Principle 21' quoted above (Article 1; cf. also Article 4.b). Moreover, the principle provides the foundation for at least *two* regional treaties: (a) the 1974 Nordic Environmental Protection Convention (UNTS 16770); and (b) the 1979 European Long-Range Transboundary Air Pollution Convention (UNTS 21623). The notion has also provided the basis for resolving a number of intergovernmental disputes via bilaterally binding international court or arbitral decisions. It has additionally been reiterated in some non-binding United Nations resolutions noted below.

Thus it is crucial to recognize that it is societal attitudes that ultimately determine the level of protection afforded the environment, whether in peacetime or wartime. Indeed, the importance of cultural norms in determining military actions is fully realized and firmly imbedded in the Law of War: *those military actions not precisely regulated are to be controlled by the principles of humanity and the dictates of the public conscience.* This so-called Martens principle of international law is first found in 1899 Hague Convention II through the efforts of Feodor F. Martens [1845–1909] and derives from 'the usages established between civilized nations, from the laws of humanity, and the requirements of the public conscience' (Preamble). This praiseworthy notion has been in essence reiterated in a considerable number of subsequent multilateral treaties within the Law of War.

The evolving dictates of the public conscience are to some considerable extent reflected in the hortatory declarations made by the United Nations, these often following lengthy debate and detailed roll-call voting. To that end it is instructive to note the strong support that was given to a number of key pronouncements. Thus, in 1972 the United Nations Declaration on the Human Environment proclaimed not only that states have the responsibility to ensure that activities within their control do not cause damage to the environment beyond their own borders (UNGA, 1972b, Principle 21); but, moreover, that humans and their environment must be spared the effects of nuclear weapons and all other means of mass destruction (UNGA, 1972b, Principle 26). The 1982 United Nations World Charter for Nature proclaimed that nature shall be secured against degradation caused by warfare (UNGA, 1982, Article 5); and, moreover, that areas degraded by human activities shall be rehabilitated (UNGA, 1982, Article 11.e). And the 1992 United Nations Declaration on Environment and Development proclaimed that states shall respect international law providing protection for the environment in times of warfare (UNGA, 1992b, Principle 24). Not all such resolutions are as widely endorsed as were those just presented. One that unfortunately did not fare so well in 1980 proclaimed that states have the responsibility to preserve their own environment for present and future generations, at the same time drawing attention to the pernicious effects of military expenditures on the environment (UNGA, 1980). In short, these aspirational statements, although not of a binding nature, do suggest the emergence and strength of cultural norms; and at the same time they contribute to the progressive development of legal norms.

Perhaps the most valuable approach to strengthening the legal regime protective of the environment in time of war would be to foster *environmental education*, both formal and informal and focusing such attention on both the civil and military sectors of society (Westing, 1988). One largely overlooked aspect of the relevant environmental education would be to teach both the armed forces and the public at large the substance of the relevant treaties to which a state has committed itself. In fact, such dissemination has actually been mandated in a number of important instances. An imperative corollary to such dissemination within the military sector would be for the armed forces of the world to revise their military manuals as appropriate to reflect environmental concerns (Westing, 2000).

Various intergovernmental agencies and nongovernmental organizations—including especially the *United Nations Educational, Scientific and Cultural Organization* (Paris), the *United Nations Environment Programme*, the *International Committee of the Red Cross* (Geneva), and the *International Union for Conservation of Nature*—could take leading roles in facilitating the necessary environmental education in the many states where it is currently inadequate. Within the civil sector, specialized attention should be given to schools of law, of journalism, and of divinity.

The applicability of several realms of international law to environmental disruption in time of war was noted earlier. This inevitably leads to some overlap, which is all to the good. For example, it is of some interest to stress that protection of the environment in time of war has become an unequivocal component of International Humanitarian Law (especially via 1977 Bern Protocol I, Articles 35 & 55). As a result, the *International Committee of the Red Cross*, the universally respected guardian of that realm of the law, has been expanding its humanitarian concerns to include the environment. Similarly, the *United Nations Environment Programme* has to some extent at least become involved in environmental matters as these relate to wartime disruption. Such linkages should be encouraged to insure that environmental concerns be properly considered within the context of all relevant human actions.

5.6 Conclusion

The demands upon the global environment and its resources that emanate from the civil sector of society are of such magnitude that every effort must be made to avoid unnecessary utilization or damage from any direction. Warfare provides a stark example of unnecessary, avoidable environmental utilization and disruption. Military disruption of the environment is pernicious because it spills over both the spatial and temporal bounds of the attack, because it has partially unpredictable ramifications, and because its impact assails combatants and non-combatants alike. The Gulf War provides an especially poignant example of unacceptable behavior, not only because of its appalling human costs, but additionally because the human environment was substantially degraded in a region that can ill afford such loss as it strives to achieve sustainable development.

It remains the task of the international community to render assistance in ameliorating those social and environmental impacts of the Gulf War that continue to undermine the ability of the embroiled civilian populations to survive. However, the primary continuing task of the international community is to develop and foster cultural norms, and accompanying legal instruments, that would serve to prevent recurrences of such calamities. And it should be primarily through a strengthened United Nations system that these goals are realized.

Multilateral treaties are in place that could have prevented the social and environmental tragedies of the Gulf War. But they are insufficiently acceded to, insufficiently complied with and enforced, and insufficiently known, understood, and supported by the public at large. On the other hand, public responses to the Gulf War have made it clear that the dictates of the public conscience have evolved to the point where wartime vandalism of the environment is no longer acceptable. Such public recognition suggests one central lesson of the Gulf War to be that the existing protective treaties remain inadequate in the absence of a public that is aware of their existence, supportive of their provisions, and ready to have their compliance mechanisms strengthened.

In closing, a strengthening of international environmental cultural and legal norms as suggested here would provide at least three benefits of exceptional importance: (a) it would provide a greater level of protection to nature; (b) it would help to erode the still largely sacrosanct status of the military sector (*i.e.*, diminish the sovereign immunity it now enjoys so widely); and (**c**) it would represent a small step towards a universal rule of law. But in the last analysis, it must never be lost sight of that the only way to avoid the terrible depredations of war, both social and environmental, is to turn to non-violent means of conflict resolution.

References

Abbott, A. 2001. WHO plans study of Gulf War fallout. *Nature* (London) 413(6852):97.

Abdulraheem, M.Y. 2000. War-related damage to the marine environment in the ROPME sea area. In: Austin, J.E., & Bruch, C.E. (eds). *Environmental Consequences of War: Legal, Economic, and Scientific Perspectives*. Cambridge, UK: Cambridge University Press, 691 pp: pp 338–352 (Chap. 13).

Ahtisaari, M., *et al.* 1991a. *Report to the Secretary-General on Humanitarian Needs in Kuwait and Iraq in the Immediate Post-crisis Environment by a Mission to the Area*. New York: *United Nations Security Council*, Document No. S/22366 (20 Mar 91), 13 pp.

Ahtisaari, M., *et al.* 1991b. *Report to the Secretary-General on Humanitarian Needs in Kuwait in the Immediate Post-crisis Environment by a Mission to the Area*. New York: *United Nations Security Council*, Document No. S/22409 (28 Mar 91), 13 pp.

Al-Awadi, A.A. 1995. *The environmental Catastrophe in Kuwait*. Safat, Kuwait: Regional Organization for the Protection of the Marine Environment [ROPME], 36 pp.

Arkin, W.M., Durrant, D., & Cherni, M. 1991. *On Impact: Modern Warfare and the Environment: a Case Study of the Gulf War*. Washington: Greenpeace, 171+8+23 pp.

Ascherio, A., et al. 1992. Effect of the Gulf War on infant and child mortality in Iraq. *New England Journal of Medicine* (Boston) 327(13):931–936.

Bakan, S., et al. 1991. Climate response to smoke from the burning oil wells in Kuwait. *Nature* (London) 351(6325):367–371.

Barnaby, F. 1991. Environmental impact of the Gulf War. *Ecologist* (Sturminster, UK) 21(4):166–172.

Beaumont, P. 1978. Euphrates River: an international problem of water resources development. *Environmental Conservation* (Cambridge, UK) 5(1):35–43.

Birchard, K. 1998. Does Iraq's depleted uranium pose a health risk? *Lancet* (London) 351(9103):657.

Boutros-Ghali, B. 1992. *Protection of the Environment in Times of Armed Conflict*. New York: *United Nations General Assembly*, Document No. A/47/328 (31 Jul 92), 15 pp.

Bouvier, A. 1991. Protection of the natural environment in time of armed conflict. *International Review of the Red Cross* (Geneva) 31(285):567–578.

Brown, G., & Porembski, S. 2000. Phytogenic hillocks and blow-outs as 'safe sites' for plants in an oil-contaminated area of northern Kuwait. *Environmental Conservation* (Cambridge, UK) 27(3):242–249.

Browning, K.A., et al. 1991. Environmental effects from burning oil wells in Kuwait. *Nature* (London) 351(6325):363–367.

Canby, T.Y. 1991. After the storm. *National Geographic* (Washington) 180(2):2–33.

Casey-Vine, P. 1991. Legal aspects of environmental warfare. In: McKinnon, M., & Vine, P. *Tides of War*. London: Boxtree, 192 pp: pp 181–185.

Charrier, B. 1998. *Environmental Assessment of Kuwait Seven Years after the Gulf War: Final Report*. Geneva: Green Cross International, 4 pp.

CIA. 1992. *World Factbook 1992*. Washington: US Central Intelligence Agency, 439 pp + 16 maps.

Cloudsley-Thompson, J.L. 1991. Environmental impact of the Gulf War. *Environmental Awareness* (Vadodara, India) 14(1):9–10.

Collins, J.M. 1990. *Military Geography of Iraq and adjacent Arab Territory*. Washington: US Library of Congress, Congressional Research Service, Report for Congress No. 90–431, 15 pp.

Earle, S.A. 1992. Persian Gulf pollution: assessing the damage one year later. *National Geographic* (Washington) 181(2):122–134.

Eglin, D.R., & Rudolph, J.D. 1985. Kuwait. In: Nyrop, R.F. (ed.). *Persian Gulf States: Country Studies*. 2nd edn. Washington: US Department of the Army, Pamphlet No. 550-185, 540 pp: pp 73–141.

Ekéus, R. 1992. United Nations special commission on Iraq. *SIPRI Yearbook* (Oxford) 1992: 509–530.

Falk, R.[A.]. 1991. Questioning the UN mandate in the Gulf. *IFDA [International Federation for Development Alternatives] Dossier* (Nyon, Switzerland) 1991(81):81–88.

Fauteux, P. 1991. Environmental law and the Gulf War. *IUCN [International Union for Conservation of Nature] Bulletin* (Gland, Switzerland) 22(3):26–27.

Fauteux, P. 1992a. The Gulf War, the ENMOD Convention and the Review conference. *UNIDIR [United Nations Institute for Disarmament Research] Newsletter* (Geneva) 1992(18):6–12.

Fauteux, P. 1992b. Use of the environment as an instrument of war in occupied Kuwait. In: Schiefer, H.B. (ed.). *Verifying Obligations Respecting Arms Control and the Environment: a Post Gulf War Assessment*. Saskatoon, Canada: University of Saskatchewan, 231 pp: pp 35–79.

Feinstein, L. 1991. Iraqi nuclear, chemical, and biological facilities attacked. *Arms Control Today* (Washington) 21(2):19–20.

Ffrench, G.E., & Hill, A.G. 1971. *Kuwait: Urban and Medical Ecology: a Geomedical Study*. Berlin: Springer-Verlag, 124 pp + 3 maps.

Fotion, N.G. 1991. Gulf War: cleanly fought. *Bulletin of the Atomic Scientists* (Chicago) 47(7):24–29.

Geiger, H.J. 1994. Bomb now, die later: the consequences of infrastructure destruction for Iraqi civilians in the Gulf War. In: O'Loughlin, J., Mayer, T., & Greenberg, E.S. (eds). *War and its Consequences: Lessons from the Persian Gulf Conflict*. New York: HarperCollins, 252 pp: pp 51–58 + 236–252 *passim*.

Goldblat, J. 1991. Legal protection of the environment against the effects of military activities. *Bulletin of Peace Proposals* [now *Security Dialogue*] (Oslo) 22(4):399–406.

Green, L.C. 1991. Environment and the law of conventional warfare. *Canadian Yearbook of International Law* (Vancouver) 29:222–237.

Greenpeace. 1992. *Environmental Legacy of the Gulf War*. Amsterdam: Greenpeace International, 42 pp.

Gulf Task Force, US. n.d. [1991 or 1992]. *Environmental Crisis in the Gulf: the U.S. Response*. Washington: President of the United States, 20 pp.

Hawley, T.M. 1992. *Against the Fires of Hell: the Environmental Disaster of the Gulf War*. New York: Harcourt Brace Jovanovich, 208 pp.

Hegazy, A.K. 1997. Plant succession and its optimization on tar-polluted coasts in the Arabian Gulf region. *Environmental Conservation* (Cambridge, UK) 24(2):149–158.

Heinebäck, B. 1974. *Oil and Security*. Stockholm: Almqvist & Wiksell, 197 pp.

Hobbs, P.V., & Radke, L.F. 1992. Airborne studies of the smoke from the Kuwait oil fires. *Science* (Washington) 256(5059):987–991; 259(5103):1811–1812.

Horváth, G., & Zell, J. 1996. Kuwait oil lakes as insect traps. *Nature* (London) 379(6563): 303–304.

IUCN. 1985. *Management and Conservation of Renewable Marine Resources in the Kuwait Action Plan Region*. Nairobi: *United Nations Environment Programme*, Regional Seas Reports & Studies No. 63, 57 pp.

IUCN. 1990a. *1990 IUCN Red List of Threatened Animals*. Gland, Switzerland: *International Union for Conservation of Nature (IUCN)*, 192 pp.

IUCN. 1990b. *1990 United Nations List of National Parks and Protected Areas*. Gland, Switzerland: *International Union for Conservation of Nature (IUCN)*, 275 pp.

Johnson, D.W., *et al.* 1991. Airborne observations of the physical and chemical characteristics of the Kuwait oil smoke plume. *Nature* (London) 353(6345):617–621.

Joffe, A.H. 2000. Environmental legacy of Saddam Husayn: the archaeology of totalitarianism in modern Iraq. *Crime, Law & Social Change* (Dordrecht, Netherlands) 33(4):313–328.

Kapp, C. 2001. WHO sends team to Iraq to investigate effects of depleted uranium. *Lancet* (London) 358(9283):737.

Karrar, G., Batanouny, K.H., Mian, M.A., & El-Din, M.N.A. 1991. *Rapid Assessment of the Impacts of the Iraq-Kuwait Conflict on Terrestrial Ecosystems. I. The Republic of Iraq. II. The State of Kuwait. III. The Kingdom of Saudi Arabia*. Nairobi: *United Nations Environment Programme*, Terrestrial Ecosystems Branch, 3 vols (77 + 87 + 67 pp + apps) (Sep 91).

Karsh, E. (ed.). 1989. *Iran–Iraq War: Impact and Implications*. London: Macmillan, 303 pp.

Kaufmann, J., Leurdijk, D., & Schrijver, N. 1991. *World in Turmoil: Testing the UN's Capacity*. Hanover, NH, USA: Academic Council on the United Nations System, Reports & Papers No. 1991–4, 150 pp.

Khuraibet, A.M. 1999. Nine years after the invasion of Kuwait: the impacts of the Iraqi left-over ordnance. *Environmentalist* (Dordrecht, Netherlands) 19(4):361–368.

Lawler, A. 2001. Destruction in Mesopotamia. *Science* (Washington) 293(5527):32–35.

Lee, I., & Haines, A. 1991. Health costs of the Gulf War. *British Medical Journal* (London) 303(6797):303–306.

Lee, M.R. (ed.). 1992. *Environmental Aftermath of the Gulf War*. Washington: US Senate, Committee on Environment & Public Works, Committee Print No. S.Prt.102–84, 74 pp.

Lewis, P. 2001. From UNSCOM to UNMOVIC: the United Nations and Iraq. *Disarmament Forum* (Geneva) 2001(2):63–68.

Limaye, S.S., Suomi, V.E., Velden, C., & Tripoli, G. 1991. Satellite observations of smoke from oil fires in Kuwait. *Science* (Washington) 252(5012):1536–1539.

Lindén, O., *et al.* 1990. *State of the Marine Environment in the ROPME Sea Area*. 1st rev. Nairobi: *United Nations Environment Programme*, Regional Seas Reports & Studies No. 112, 34 pp.

Lopez, G.A. 1991. Gulf War: not so clean. *Bulletin of the Atomic Scientists* (Chicago) 47(7):30–35.

Luttwak, E.N. 1994. The Gulf War in its purely military dimension. In: O'Loughlin, J., Mayer, T., & Greenberg, E.S. (eds). *War and its Consequences: Lessons from the Persian Gulf Conflict*. New York: HarperCollins, 252 pp: pp 33–50 + 236–252 *passim*.

McKinnon, M., & Vine, P. 1991. *Tides of War*. London: Boxtree, 192 pp.
Metz, H.C. (ed.). 1990. *Iraq: a Country Study*. 4th edn. Washington: US Department of the Army, Pamphlet No. 550–31, 299 pp.
Michel, P.-F. (ed.). 1991. *The Gulf 1990–1991: from Crisis to Conflict: the ICRC at Work*. Geneva: International Committee of the Red Cross, 48 pp.
Oliver, F.W. 1945–1946. Dust-storms in Egypt and their relation to the war period, as noted in Maryut, 1939–45. *Geographical Journal* (London) 106(1–2):26–49 + 4 pl.; 108(4–6):221–226 + 1 pl.
Omar, S.A.S., Briskey, E., Misak, R., & Asem, A.A.S.O. 2000. Gulf War impact on the terrestrial environment of Kuwait: an overview. In: Austin, J.E., & Bruch, C.E. (eds). *Environmental Consequences of War: Legal, Economic, and Scientific Perspectives*. Cambridge, UK: Cambridge University Press, 691 pp: pp 316–337 (Chap. 12).
Oza, G.M. 1991. Gulf War: environmental disaster for wildlife and mankind. *Environmental Awareness* (Vadodara, India) 14(1):1–7.
Pawlick, T. 1991. Operation boogar man. *National Wildlife* (Vienna, VA, USA) 29(5):29.
Plant, G. (ed.). 1992. *Environmental Protection and the Law of War: a 'Fifth Geneva' Convention on the Protection of the Environment in Time of Armed Conflict*. London: Belhaven Press, 284 pp.
Pope, C. 1991. War on earth. *Sierra* (San Francisco) 76(3):54–58.
Price, A. 1991. Gulf conflict: oil on troubled waters. *Environmental Awareness* (Vadodara, India) 14(1):11–13.
Readman, J.W., *et al.* 1992. Oil and combustion-product contamination of the Gulf marine environment following the war. *Nature* (London) 358(6388):662–665.
Renner, M.G. 1991. Military victory, ecological defeat. *World-Watch* (Washington) 4(4):27–33.
Roberts, A. 1993. Failures in protecting the environment in the 1991 Gulf war. In: Rowe, P. (ed.). *Gulf War 1990-91 in International and English Law*. London: Routledge, 463 pp: pp 111–154.
Robinson, N.A. 1991. International law and the destruction of nature in the Gulf War. *Environmental Policy & Law* (Bonn) 21(5–6):216–220.
Sandoz, Y. 1992. Protection of the environment in time of war. *UNIDIR [United Nations Institute for Disarmament Research] Newsletter* (Geneva) 1992(18):12–14.
Schachter, O. 1991. United Nations law in the Gulf conflict. *American Journal of International Law* (Washington) 85(3):452–473.
Seager, J. 1992. Operation desert disaster: environmental costs of the war. In: Peters, C. (ed.). *Collateral Damage: the New World Order at Home and Abroad*. Boston: South End Press, 433 pp: pp 197–216.
Sheppard, C., & Price, A. 1991. Will marine life survive the Gulf War? *New Scientist* (London) 129(1759):36–40.
Shouse, B. 2001. Gulf War's aftermath: Kuwait unveils plan to treat festering desert wound. *Science* (Washington) 293(5534):1410.
Small, R.D. 1991. Environmental impact of fires in Kuwait. *Nature* (London) 350(6313):11–12.
Terry, J.P. 1992. Environment and the laws of war: the impact of Desert Storm. *Naval War College Review* (Newport, RI, USA) 45(1):61–67.
Tolba, M.K. 1991. *Environmental Consequences of the Armed Conflict between Iraq and Kuwait*. Nairobi: *United Nations Environment Programme*, Document No. UNEP/GC.16/4/Add.1 (10 May 91), 10 pp.
Trux, J. 1991. Desert Storm: space-age war. *New Scientist* (London) 131(1779):30–34.
UNEP. 1990. *Situation in the Middle East*. Nairobi: *United Nations Environment Programme*, Decision No. SS.II/8 (3 Aug 90), 1 p.
UNEP. 1991. *Military Conflicts and the Environment: Environmental Consequences of the Armed Conflict in the Gulf Area*. Nairobi: *United Nations Environment Programme*, Decision No. 16/11.A (31 May 91), 1 p.
UNEP. 1991–1992. *Report on the UN Inter-agency Plan of Action for the ROPME Region: Phase I. Initial Surveys and Preliminary Assessment*. Nairobi: *United Nations Environment*

Programme, Oceans & Coastal Areas Programme Activity Centre, 48 + 6 pp (12 Oct 91 & 12 Jun 92).

UNGA. 1972a. *Co-operation between States in the Field of the Environment*. New York: *United Nations General Assembly*, Resolution No. 2995(XXVII) (15 Dec 72), 1 p. [112 (85 %) in favor, 10 abstentions, 0 against, 10 absent = 132]

UNGA. 1972b. Declaration of the United Nations conference on the human environment. In: *UNGA*. 1973. *Report of the United Nations Conference on the Human Environment, Stockholm, 5–16 June 1972*. New York: *United Nations General Assembly*, Document No. A/CONF.48/14/Rev.1, 77 pp: pp 3–5.

UNGA. 1980. *Historical Responsibility of States for the Preservation of Nature for Present and Future Generations*. New York: *United Nations General Assembly*, Resolution No. 35/8 (30 Oct 80), 1 p. [70 (45 %) in favor, 47 abstentions, 0 against, 37 absent = 154]

UNGA. 1982. *World Charter for Nature*. New York: *United Nations General Assembly*, Resolution No. 3/77 (28 Oct 82), 5 pp. [114 (73 %) in favor, 17 abstentions, 1 against, 25 absent = 157]

UNGA. 1991a. *Exploitation of the Environment in Times of Armed Conflict and the Taking of Practical Measures to Prevent such Exploitation*. New York: *United Nations General Assembly*, Decision No. 46/417 (9 Dec 91), 1 p. [166 (100 %) in favor (adopted without a vote)]

UNGA. 1991b. *International Cooperation to Mitigate the Environmental Consequences on Kuwait and other Countries in the Region Resulting from the Situation between Iraq and Kuwait*. New York: *United Nations General Assembly*, Resolution No. 46/216 (20 Dec 91), 2 pp. [136 (82 %) in favor, 1 abstention, 0 against, 29 absent = 166]

UNGA. 1992a. *International Cooperation to Mitigate the Environmental Consequences on Kuwait and other Countries in the Region Resulting from the Situation between Iraq and Kuwait*. New York: *United Nations General Assembly*, Resolution No. 47/151 (18 Dec 92), 2 pp. [159 (89 %) in favor, 2 abstentions, 0 against, 18 absent = 179]

UNGA. 1992b. *Rio Declaration on Environment and Development*. New York: *United Nations General Assembly*, Document No. A/CONF.151/5/Rev.1 (13 Jun 92), 6 pp.

UNSC. 1983. *[Situation between Iran and Iraq.]* New York: *United Nations Security Council*, Resolution No. S/RES/540(1983) (31 Oct 83), 1 p. [12 (80 %) in favor, 3 abstentions, 0 against, 0 absent = 15]

UNSC. 1990a. *[Condemnation of the Iraqi Invasion of Kuwait.]* New York: *United Nations Security Council*, Resolution No. S/RES/660(1990) (2 Aug 90), 1 p. [14 (93 %) in favor, 1 abstention, 0 against, 0 absent = 15]

UNSC. 1990b. *[Implementation of Resolution No. 660(1990).]* New York: *United Nations Security Council*, Resolution No. S/RES/678(1990) (29 Nov 90), 2 pp. [12 (80 %) in favor, 1 abstention, 2 against, 0 absent = 15]

UNSC. 1991a. *[Suspension of Offensive Combat Operations.]* New York: *United Nations Security Council*, Resolution No. S/RES/686(1991) (2 Mar 91), 3 pp. [11 (73 %) in favor, 3 abstentions, 1 against, 0 absent = 15]

UNSC. 1991b. *[The Restoration to Kuwait of its Sovereignty.]* New York: *United Nations Security Council*, Resolution No. S/RES/687(1991) (3 Apr 91), 10 pp. [12 (80 %) in favor, 2 abstentions, 1 against, 0 absent = 15]

Urquhart, B. 1991. Role of the United Nations in the Iraq-Kuwait conflict in 1990. *SIPRI Yearbook* (Oxford) 1991:617–637.

Walker, P.F., & Stambler, E. 1991….and the dirty little weapons. *Bulletin of the Atomic Scientists* (Chicago) 47(4):20–24.

Warner, F. 1991. Environmental consequences of the Gulf War. *Environment* (Washington) 33(5):6–9,25–26.

Westing, A.H. 1976. *Ecological Consequences of the Second Indochina War*. Stockholm: Almqvist & Wiksell, 119 pp + 8 pl.

Westing, A.H. 1977. *Weapons of Mass Destruction and the Environment*. London: Taylor & Francis, 95 pp.

Westing, A.H. 1980. *Warfare in a Fragile World: Military Impact on the Human Environment.* London: Taylor & Francis, 249 pp.

Westing, A.H. (ed.). 1985a. *Explosive Remnants of war: Mitigating the Environmental Effects.* London: Taylor & Francis, 141 pp.

Westing, A.H. 1985b. Misspent energy: munition expenditures past and future. *Bulletin of Peace Proposals* [now *Security Dialogue*] (Oslo) 16(1):9–10.

Westing, A.H. (ed.). 1988. *Cultural Norms, War and the Environment.* Oxford: Oxford University Press, 177 pp.

Westing, A.H. 1989. Herbicides in warfare: the case of Indochina. In: Bourdeau, P., *et al.* (eds). *Ecotoxicology and Climate: with Special Reference to Hot and Cold Climates.* Chichester, UK: John Wiley, 392 pp: pp 337–357.

Westing, A.H. (ed.). 1990a. *Environmental Hazards of War: Releasing Dangerous Forces in an Industrialized World.* London: Sage Publications, 96 pp.

Westing, A.H. 1990b. Towards eliminating war as an instrument of foreign policy. *Bulletin of Peace Proposals* [now *Security Dialogue*] (Oslo) 21(1):29–35.

Westing, A.H. 1994. Constraints on environmental disruption during the Gulf War. In: O'Loughlin, J., Mayer, T., & Greenberg, E.S. (eds). *War and its Consequences: Lessons from the Persian Gulf Conflict.* New York: HarperCollins, 252 pp: pp 77–84 + 236–252 *passim.*

Westing, A.H. 1996. Explosive remnants of war and the human environment. *Environmental Conservation* (Cambridge, UK) 23(4):283–285.

Westing, A.H. 1997. Environmental protection from wartime damage: the role of international law. In: Gleditsch, N.P. (ed.). *Conflict and the Environment.* Dordrecht, Netherlands: Kluwer Academic Publishers, 598 pp: pp 535–553.

Westing, A.H. 2000. In furtherance of environmental guidelines for armed forces during peace and war. In: Austin, J.E., & Bruch, C.E. (eds). *Environmental Consequences of War: Legal, Economic, and Scientific Perspectives.* Cambridge, UK: Cambridge University Press, 691 pp: pp 171–181 (Chap. 6).

Weston, B.H. 1991. Security Council Resolution 678 and Persian Gulf decision making: precarious legitimacy. *American Journal of International Law* (Washington) 85(1):516–535.

White, M. 1991. Environmental consequences of the Gulf War. *Environmental Awareness* (Vadodara, India)) 14(1):14–16.

Wolkomir, R., & Wolkomir, J. 1992. Caught in the crossfire. *International Wildlife* (Vienna, VA, USA) 22(1):4–11.

Yergin, D. 1991. *The prize: the Epic Quest for Oil, Money, and Power.* New York: Simon & Schuster, 876 + xxxiii pp + 32 pl.

York, C. 1991. International law and the collateral effects of war on the environment: the Persian Gulf. *South African Journal on Human Rights* (Johannesburg) 7(3):269–290.

Zedalis, R.J. 1991. Burning of the Kuwaiti oilfields and the laws of war. *Vanderbilt Journal of Transnational Law* (Nashville, TN, USA) 24(4):711–755.

Zilinskas, R.A. 1997. Iraq's biological weapons: the past as future? *Journal of the American Medical Association* (Chicago) 278(5):418–424.

Zimansky, P., & Stone, E.C. 1992a. Antiquities in the aftermath: a winter trip to Iraq. *Mar Sipri* (Boston) 5(1):1,5–8.

Zimansky, P., & Stone, E.C. 1992b. Mesopotamia in the aftermath of the Gulf War. *Archaeology* (New York) 45(3):24.

Chapter 6
Environmental War: Hostile Manipulations of the Environment

Note *: The pursuit of war is almost unavoidably damaging to the environment, particularly so, of course, within its theater of operations (#108, #139, #204, #243, #304).*[1] *Early on in my studies of the multifarious environmental impacts of the Second Indochina War of 1961–1975* (cf. *Chap. 4), I recognized that* three *levels of hostile environmental damage could be distinguished: (a) Unintentional damage (also referred to as Collateral damage); (b) Intentional damage; and (c) Amplified damage (also referred to as Environmental warfare). It is this last category of environmental war—one in which a relatively modest expenditure of triggering energy leads to a substantially greater amount of destructive energy—that is the subject of this chapter. Moreover, it is an approach to warfare that became the stimulus for the 1977 Convention on the Prohibition of Military or any other Hostile Use of Environmental Modification Techniques (UNTS 17119), a specific extension of the Law of War, albeit in my view a rather weak and unsatisfactory one (#234). Finally, from what is reproduced below (#144) it may be of interest to learn that perhaps the most devastating single action in all human history was an act of environmental war (as defined here), one that was perpetrated in China in 1938.*[2]

---

[1] The numbered references are provided in Chap. 3.

[2] Reproduced from: Westing, A.H. (ed.). *Environmental Warfare: a Technical, Legal and Policy Appraisal*. London: Taylor & Francis, 107 pp: pp 3–12 (Chap. 1); 1984, with the original title "*Environmental Warfare—An Overview*", by permission of the *Stockholm International Peace Research Institute (SIPRI)*, the copyright holder, on 20 March 2012.

A. H. Westing, *Arthur H. Westing*, SpringerBriefs on Pioneers in Science and Practice 1, DOI: 10.1007/978-3-642-31322-6_6,

6.1 Introduction

Environmental warfare refers to the manipulation of the environment for hostile military purposes. The militarily most useful hostile manipulations of the environment would be those in which a relatively modest expenditure of triggering energy leads to the release of a substantially greater amount of directed destructive energy.

Environmental warfare could, at least in principle, involve damage-causing manipulations of: (a) celestial bodies or space; (b) the atmosphere; (c) the land (lithosphere); (d) the oceans (hydrosphere); or (e) the biota, either terrestrial or marine (biosphere). Each of these five domains is considered in turn in the sections that follow.

A number of the hostile manipulations of today and tomorrow that comprise environmental warfare fall under the aegis of a number of disparate arms control treaties, either directly or indirectly. These legal restraints are alluded to in the concluding section. Prominent among them is the 1977 Environmental Modification (Enmod) Convention (*UNTS* 17119) (cf. also Goldblat, 1984; Krass, 1984). Policy recommendations are also given separate treatment (Westing *et al.*, 1984).

This analysis draws to some extent upon two earlier works by the author (Westing, 1977; 1980). Moreover, a catalogue is available elsewhere of potential hostile manipulations of the environment (Canada, 1975).

6.2 Celestial Bodies and Space

Celestial bodies refer to the Moon and other such planetary satellites, the planets themselves, the Sun and other stars, asteroids, meteors, and the like. *Space* refers to all of the vast region beyond our atmosphere (*i.e.*, the region above the ionosphere) and thus, for practical purposes, begins a few hundred kilometers above the Earth's surface.

With reference to the hostile manipulation of celestial bodies, it was suggested recently that some day we might have the ability to divert asteroids, using a nuclear weapon, so as to cause them to strike enemy territory (Sullivan, 1983). There appears to be no suggestion as yet for how space might be manipulated for hostile purposes.

6.3 The Atmosphere

The Earth's atmosphere extends upwards some hundreds of kilometers, but becomes extraordinarily thin beyond ca 150 km. It is divided into the lower atmosphere, which represents more than 99 % of the atmospheric mass, and the upper atmosphere, with less than 1 % of the mass. The lower atmosphere extends upward perhaps 55 km. It consists of the troposphere (ca 0–12 km up; ca 87 % of the total atmospheric mass) and the stratosphere (ca 12–55 km). The stratosphere, in turn, can be divided into the lower stratosphere (ca 12–30 km) and the upper stratosphere (ca 30–55 km). The troposphere is turbulent (windy) and contains clouds whereas the stratosphere is quiescent and cloudless. The upper atmosphere (above ca 55 km) consists of the mesosphere (ca 55–80 km) and the ionosphere (or thermosphere) (ca 80–a few hundred km). The ionized (electrified) molecules that distinguish the ionosphere serve to deflect certain radio waves downwards, thereby making possible long-distance amplitude-modulated (AM) radio communication.

Some ozone is found throughout the atmosphere, its overall average concentration being 635 μg/kg (820/$\mu g/m^3$). The atmospheric ozone is not distributed evenly, but is found largely in the lower stratosphere, indeed, largely within a so-called ozone layer (ca 20–30 km up) in which the atmospheric concentration of ozone is up to 100 times the overall average. This ozone layer provides a partial barrier to solar ultraviolet radiation.

With respect to hostile manipulations of the upper atmosphere, it is sometimes suggested that techniques might be developed in the future which would make it possible to alter the electrical properties of the ionosphere in such a way as to disrupt enemy communications.[3] In fact, during the early 1960s the US Air Force carried out some short-lived experiments in which huge numbers of tiny lengths of fine copper wire were injected into the ionosphere—in this instance, however, for the opposite purpose of improving radio communications (Liller, 1964; Stevenson, 1963).

With respect to the lower atmosphere, some consider it an imminent possibility to be able to open a temporary 'window' in the ozone layer above enemy territory for the purpose of permitting an injurious level of ultraviolet radiation to penetrate to the ground, perhaps by the controlled release of a bromine compound from orbiting satellites (Sullivan, 1975). However, the direct military utility of such an action, even if it could be accomplished, would seem to be exceedingly low.

There is a report that the USA injected unknown substances into the troposphere over enemy territory during the Second Indochina War of 1961–1975 for the purpose of rendering inoperable enemy radars used for aiming defensive surface-to-air missiles (Hersh, 1972). This operation has not been acknowledged and, if it indeed occurred, there is no indication of the extent to which it succeeded.

Various levels of control over winds (*e.g.*, creation or redirection of hurricanes), over clouds (*e.g.*, creation or dissolution of fog, generation of cloud-to-ground lightning), or over precipitation (*e.g.*, production of torrential rains, heavy snowfall, massive hail) could bring about direct or indirect damage to an enemy. The

[3] High-altitude nuclear detonations would disrupt communication systems on the ground, but would do so directly by emitting an electro-magnetic pulse and not via an atmospheric manipulation (Stein, 1983).

effective control of winds still remains beyond human capability. Control over clouds for hostile (or other) purposes remains to date at the nonexistent to trivial levels (Atlas, 1977; Kerr, 1982; Mason, 1980). The one vigorously sustained attempt at rain making for hostile purposes—that by the USA during the Second Indochina War—achieved only indifferent, if any, success, technical or military (Westing, 1977, pp 55–57). A more detailed analysis of present and future capabilities regarding atmospheric manipulations is provided elsewhere (cf. Mészáros, 1984; cf. also Krass, 1984, pp 77–80).

A large-scale nuclear war would, of course, be extraordinarily disruptive to the human environment (Westing, 1981; 1982). Recent theoretical examinations of the subject have suggested that such an event would have an especially deleterious impact on the weather (Covey *et al.*, 1984; Turco *et al.*, 1983) and thus, in turn, on the biota (Ehrlich *et al.*, 1983). This non-directed, collateral impact of nuclear war on the atmosphere—often referred to as the 'nuclear winter'—would, it is suggested, seriously affect an area perhaps as large as half the globe for a period of weeks or months.

6.4 The Lithosphere

Land covers almost 15 thousand million hectares (29 %) of the Earth's surface. Almost 1.6 thousand million hectares of the land (11 % of the total land area) is continuously ice-covered, much of this represented by Antarctica. Perhaps 1800 million hectares (12 %) is desert. On another 800 million hectares (5 %) at least some stratum of the soil remains frozen the year round, a condition referred to as permafrost; and 200 million hectares or more (1.5 %) is accounted for by rugged mountain terrain. Much of the remaining 10,500 million hectares or so (71 %) of the land is found largely in the northern hemisphere and supports virtually the entire global population and its artifacts.

Successful manipulation of the land for hostile purposes would depend for the most part upon the ability to recognize and take advantage of local instabilities or pent-up energies, whether natural or anthropogenic. For example, some mountainous landforms are at least at certain times prone to landslides (soil and rock avalanches), and some arctic or alpine sites can be prone to snow avalanches; under the right conditions either could be initiated with hostile intent. The hostile manipulation of permafrost is taken up in Section 6.6 below. A number of important rivers flow through more than one country. This situation can provide the opportunity for an upstream nation to divert the waters of such a river so as to deny their use to a downstream nation. Natural levees or constructed dikes and dams (semi-permanent anthropogenic additions to the environment) could be destroyed to release the water contained behind them; and nuclear power stations or related facilities (further cultural artifacts that have become semi-permanent features of the environment) could be damaged so as to release their radioactive contents to the surroundings. More fanciful possibilities have also been mentioned, including the instigation of earthquakes in enemy territory or the awakening of similarly located quiescent volcanoes. A more detailed analysis of present and future capabilities regarding geospheric (tectonic) manipulations is provided elsewhere (Noltimier, 1984).

At certain times and places, appropriate military actions can bring about highly destructive floods. The most straightforward means of accomplishing this is to breach existing levees, dikes, or dams by one means or another. In a notable early instance, during the Franco–Dutch War of 1672–1678, the Dutch in June 1672 were partially successful in stopping the French from overrunning the Netherlands by cutting dikes to create the so-called Holland Water Line (Baxter, 1966, pp 72–73; Blok, 1907, pp 380–381). It might be added that this maneuver was carried out despite the vehement objections of the local inhabitants.

The Second Sino–Japanese War of 1937–1945 provides a far more devastating example of intentional military flooding (Westing, 1977, p. 54). In order to curtail the Japanese advance, the Chinese in June 1938 dynamited the Huayuankow dike of the Yellow River (Huang He) near Chengchow. This action resulted in the drowning of several thousand Japanese soldiers and stopped their advance into China along this front. In the process, however, the flood waters also ravaged major portions of Henan, Anhui and Jiangsu provinces. Several million hectares of farmlands were inundated in the process, and the crops and topsoil destroyed. The river was not brought back under control until 1947. In terms of more direct human impact, the flooding inundated some 11 Chinese cities and more than 4000 villages. At least several hundred thousand Chinese drowned as a result (and possibly many more) and several million were left homeless. Indeed, this act of environmental warfare appears to have been the most devastating single act in all human history, in terms of numbers of lives claimed.

During World War II, the British in May 1943 destroyed two major dams in the Ruhr valley, the Möhne and Eder (Brickhill, 1951, pp 95–108). There was a vast amount of damage: 125 factories were destroyed or badly damaged, 25 bridges vanished and 21 more were badly damaged, some power stations were destroyed, numerous coal mines were flooded, and railway lines were disrupted. Some 6500 cattle and pigs were lost and 3000 hectares of arable land was ruined. The official German figure for human losses was 1294. British Air Force authorities were enormously pleased with the results, summarized as 'maximum effect with minimum effort' (Brickhill, 1951, pp 9, 11). Also in World War II, German forces in 1944 intentionally flooded with salt water some 200,000 hectares of agricultural lands in the Netherlands (Aartsen, 1946); these lands subsequently required a huge rehabilitation programme (Dorsman, 1947).

During the Korean War of 1950–1953, US forces pursued a policy of attacking dams in North Korea (Rees, 1964, pp 381–382). The destruction of irrigation dams was considered by the USA to be among the most successful of its air operations of the Korean War (Futrell *et al.*, 1961, pp 627–628, 637).

Throughout the world there exist some 72 dams in 21 different countries that impound at least 1 thousand million cubic meters of water each, more than half of these being in either the USA or the USSR (Lane, 1984, p. 137). Indeed, six of them impound more than 100 thousand million cubic meters each. These 72 major dams, as well as scores of lesser dams and various major rivers with levees or dikes, stand ever ready as potential environmental targets.

There are now ca 297 nuclear-powered electrical generating stations throughout the world plus a further 15 that have been shut down (IAEA, 1983).[4] These stations (with an average net capacity of 559 MW[e]) are found in 25 different countries. Eight countries contain at least 10 each of these enduring facilities; more than one-quarter of them are located in the USA. In addition to the 312 power stations just noted, there exist a number of spent-fuel reprocessing plants, nuclear bomb factories, nuclear-waste storage repositories, and perhaps other land-based facilities harboring large quantities of radioactive materials.

Should any of these nuclear facilities be bombed in time of war, the possibility exists that a considerable surrounding area—measurable in terms of thousands of hectares—would become contaminated with injurious levels of strontium-90, caesium-137 and other radioactive elements (Cooper, 1978; Fetter & Tsipis, 1981; Ramberg, 1980). Such areas would defy effective decontamination and would thus remain uninhabitable for decades. It is thus fortunate that the only such nuclear station so far to have been destroyed with hostile intent (located in Iraq) had not yet begun operation at either of the two times it was attacked (Marshall, 1980; 1981).[5]

6.5 The Hydrosphere

The oceans of the world cover 71 % of the Earth's surface and border on 139 of the 169 or so nations. Indeed, some 43 of the 139 are island nations. The high seas also constitute an important military (naval) arena in their own right.

Among the hostile ocean modifications that have been suggested as military possibilities for the future are physical or chemical manipulations that are meant to disrupt acoustic (sonar) or electromagnetic properties of the attacked waters. The purpose for such attack would be the disruption of enemy underwater communication, remote sensing, navigation, and missile-guidance systems. The hostile destruction of nuclear-powered ships or of supertankers and other ships carrying poisonous cargoes is discussed in Section 6.6 below.

Another possibility for environmental warfare involving the oceans is the generation of tsunamis (seismic sea waves) for the purpose of destroying coastal cities and other near-shore facilities. One way that has been suggested for creating

---

[4] At the end of 2010 there were 441 nuclear reactors in operation in somewhat over 200 clusters, within 29 nations.

[5] The Iraqi nuclear reactor under construction in 1981 was destroyed by Israel. In 1991 the USA attacked and destroyed an operating Iraqi nuclear reactor. And in 2007 Israel attacked and destroyed a Syrian nuclear reactor about to go on line.

a tsunami on demand is to set off one or more nuclear devices in an appropriate underwater locality (Clark, 1961; cf. also Noltimier, 1984).

6.6 The Biosphere

The land supports some 4 thousand million hectares of tree-based (forest) ecosystems, ca 3 thousand million hectares of grass-based (prairie) ecosystems, almost 1 thousand million hectares of lichen-based (tundra) ecosystems, and perhaps 1.5 thousand million hectares of crops (both annual and perennial). The oceans support huge expanses of algae, attached or floating, and the marine ecosystems based on them. These divers ecosystems are all exploited by humans, who could not survive without the continued harvesting of trees, livestock, fish, and other renewable resources (Westing, 1980). These ecosystems additionally provide us with a series of more cryptic, though equally crucial, indirect services that keep our planet habitable (Bormann, 1976; Ehrlich & Mooney, 1983; Farnsworth *et al.*, 1981; Pimentel *et al.*, 1980; Westman, 1977). It must therefore be noted with considerable concern that these ecosystems can be manipulated for hostile purposes in a number of ways, among them: (a) by applying chemical poisons; (b) by introducing exotic living organisms; (c) by incendiary means; and (d) by mechanical means.

Forests can be devastated for hostile purposes over huge areas by spraying them with herbicides (plant poisons) or other means, as was demonstrated by the USA during the Second Indochina War of 1961–1975 (Westing, 1976; 1984). At certain times and places self-propagating wild fires could be initiated which would decimate large tracts of forest. For example, a temperate-zone coniferous forest might have an above-ground dry-weight biomass of 200,000 kg/ha having an energy content of 15 MJ/kg, and thus 3 TJ/ha of more or less readily releasable energy. Killing the trees (*i.e.*, the autotrophic component) of a forest ecosystem—whether by herbicides, fire or other means—can be expected to lead to substantial damage to that system's wildlife (heterotrophic component) and also to its nutrient budget, the latter via soil erosion and nutrient dumping (loss of nutrients in solution). Substantial recovery from such unbalancing of the regional ecosystem could well take decades (Westing, 1980, pp 8–10).

Prairies can be damaged for hostile purpose in the same ways that forests can (*i.e.*, by herbicidal, incendiary, or other attack). Thus during the Second Anglo-Boer War of 1899–1902 the Boers set torch to wide areas of veldt in order to deny forage to the advancing British (Wet, 1902, p. 181). At an estimated aboveground dry-weight biomass of 10,000 kg/ha for a prairie ecosystem, this represents a catastrophic loss to that system of perhaps 100 GJ/ha of captured and stored energy plus the loss for that growing season of food, and in some instances also cover, for the indigenous wildlife.

Tundra ecosystems can also be quite readily destroyed by one means or another, with serious ramifications (Westing, 1980, pp 114–127). Under normal conditions, tundra vegetation forms an insulating layer which prevents the underlying soil from thawing too deeply during the summer and from turning into a morass. With the vegetation destroyed not only would it become virtually impossible for vehicles to traverse the area, but perhaps the potential for serious erosion would be created and the delicately balanced ecosystem would be disastrously upset for many decades.

The employment of certain biological warfare agents could, in principle, introduce exotic micro-organisms into any region on a long-term, if not permanent, basis. Such an introduction could conceivably unbalance (or adversely re-balance) the regional ecosystem to such an extent that the area became uninhabitable for an indefinite period of time. To illustrate this point in a small way, the United Kingdom used the Scottish island of Gruinard (58° N 50° W) during 1941–1942 for testing the military potential of *Bacillus anthracis*, the causative agent of anthrax (Manchee *et al.*, 1981; 1983). The island remains dangerously contaminated to this day.

The several thousand oil wells drilled into the continental shelves of the world currently account for about one-fifth of world production (Westing, 1980, p. 165). Well over 5000 oil tankers ply the oceans (ca one-quarter of all merchant shipping) and annually transport about half of world production over long distances. Whereas these tankers have an average capacity of 70,000 m^3, perhaps three dozen of them are truly supertankers, holding more than 400,000 m^3 each. Then there exist ca 330 nuclear-powered ships, mostly large submarines owned by the USA or the USSR (Moore, 1984).

Marine ecosystems could be locally disrupted with hostile intent by destroying these offshore oil wells or loaded oil tankers (and other bulk carriers of poisonous cargoes) which were sailing near the shore. Recovery from such disruption (more probable as a collateral effect than a primary one) would be likely to take at least several years (Westing, 1980, pp 165–167). Similarly, the release into the near-shore marine environment of radioactive elements through the destruction of a nuclear-powered ship—while it would have little overt effect on the contaminated ecosystems itself—would preclude local human exploitation for many years (Westing, 1980, pp 160–163).

6.7 Conclusion

Control over the forces of nature for the achievement of military aims has been a human fantasy since the beginning of history. The ancient Greeks envied Zeus his ability to hurl thunderbolts. Moses was said to have been able to control the Red Sea in such a way as to drown the Egyptian forces that were pursuing the Israelites (*Exodus* 14:27–28). Joshua claimed to have caused the Sun to stand still so that the Israelites could consummate a battle with the Anamites (*Joshua* 10:12–13).

Those pent-up forces in the human environment that can at present be usefully released for hostile military purposes include especially: (**a**) the stored energy in the fuel represented by some forests; (**b**) the potential energy of the water held back by levees or dams; and (**c**) the decay-emitted energy of radioactive elements contained within nuclear facilities. The hostile disruption (unbalancing) of a forest, prairie, or other ecosystem could also be thought of in terms of energy, that is, as the dissipation of the complex organizational energy contained within that system.

The future could conceivably bring some measure of ability to manipulate for useful hostile purposes such forces of nature as hurricanes, earthquakes, or volcanoes (cf. Mészáros, 1984; Noltimier, 1984).

A number of international instruments enjoying varying levels of acceptance provide legal restraints against either environmental warfare per se or the means of waging it, whether feasible as yet or not (for the texts of these treaties, cf. Goldblat, 1982). Thus, the 1925 Geneva Protocol (LNTS 2138) (with 104 or more states parties) prohibits the use in war of chemical or biological agents; and the 1972 Bacteriological and Toxin Weapon Convention (*UNTS* 14860) (with 97 or more states parties) prohibits even the possession of biological or toxin agents. 1977 Protocol [I] Additional to the 1949 Geneva Conventions (UNTS 17512) (with 37 or more states parties) prohibits, with certain exceptions, attacks against the environment that would prejudice the health or survival of the population as well as attacks against works or installations containing dangerous forces, namely, dams, dikes and nuclear electrical generating stations. This Protocol includes a general prohibition of the use of methods or means of warfare which are intended, or may be expected, to cause widespread, long-lasting, *and* severe damage to the natural environment.[6]

The treaty most broadly applicable to environmental warfare is the 1977 Environmental Modification Convention (UNTS 17119) (with 41 or more states parties) (Goldblat, 1984; Krass, 1984). It prohibits the hostile use of environmental modification techniques having widespread, long-lasting, *or* severe effects as the means of damage. By 'environmental modification technique' here is specifically meant any technique for changing—through the 'deliberate manipulation' of natural processes—the dynamics, composition, or structure of space or of the Earth, including its atmosphere, lithosphere, hydrosphere, and biota.

As the capabilities of our planet to avoid environmental catastrophe on the one hand and military catastrophe on the other continue to diminish, one can only hope that moral, legal, common sense, or other restraint will prevent techniques of environmental warfare of today or tomorrow from exacerbating our growing dilemma (cf. Falk, 1984). Thus, the nations of the world disregard at their peril the fifth general principle of the 1982 *World Charter for Nature* (*UNGA*, 1982) that, 'Nature shall be secured against degradation caused by warfare or other hostile activities'.

---

[6] It might be noted that the partial restrictions on the use of nuclear weapons embodied in the 1967 Outer Space Treaty (UNTS 8843) and 1971 Seabed Treaty (UNTS 13678) do not prohibit the use for possible hostile environmental modifications or other purpose in these two environmental domains.

References

Aartsen, J.P. van. 1946. Consequences of the war on agriculture in the Netherlands. *International Review of Agriculture* (Rome) 37:5S–34S, 49S–70S, 108S–123S.

Atlas, D. 1977. The paradox of hail suppression. *Science* (Washington) 195(4272):139–145.

Baxter, S.B. 1966. *William III and the Defense of European Liberty 1650–1702*. New York: Harcourt, Brace & World, 462 pp + 8 pl.

Blok, P.J. 1907. *History of the People of the Netherlands. IV. Frederick Henry, John de Witt, William III* (translated from the Dutch by Bierstadt, O.A.) New York: G.P. Putnam's Sons, 566 pp + 3 maps.

Bormann, F.H. 1976. Inseparable linkage: conservation of natural ecosystems and the conservation of fossil energy. *BioScience* (Washington) 26:754–760.

Brickhill, P. 1951. *Dam Busters*. London: Evans Brothers, 269 pp + 13 pl.

Canada. 1975. *Suggested Preliminary Approach to Considering the Possibility of Concluding a Convention on the Prohibition of Environmental Modification for Military or Other Hostile Purposes*. Geneva: Conference of the Committee on Disarmament Document No. CCD/463 (5 Aug 75), 24 + 1 pp.

Clark, W.H. 1961. Chemical and thermonuclear explosives. *Bulletin of the Atomic Scientists* (Chicago) 17:356–360.

Cooper, C.L. 1978. Nuclear hostages. *Foreign Policy* (Washington) 1978(32):127–135.

Covey, C., et al. 1984. Global atmospheric effects of massive smoke injections from a nuclear war: results from general circulation model simulations. *Nature* (London) 308:21–25.

Dorsman, C. 1947. [Damage to horticultural crops from inundation with seawater.] (In Dutch). *Tijdschrift over Plantenziekten* (Wageningen, Netherlands) 53(3):65–86.

Ehrlich, P.R., & Mooney, H.A. 1983. Extinction, substitution, and ecosystem services. *BioScience* (Washington) 33:248–254.

Ehrlich, P.R., *et al.* 1983. Long-term biological consequences of nuclear war. *Science*, (Washington) 222:1293–1300.

Falk, R.A. 1984. Environmental disruption by military means and international law. In: Westing, A.H. (ed.). *Environmental Warfare: a Technical, Legal and Policy Appraisal*. London: Taylor & Francis, 107 pp: pp 33–51 (Chap. 4).

Farnsworth, E.G., *et al.* 1981. Value of natural ecosystems: an economic and ecological framework. *Environmental Conservation* (Cambridge, UK) 8:275–282.

Fetter, S.A., & Tsipis, K. 1981. Catastrophic releases of radioactivity. *Scientific American* (New York) 244(4):33–39, 146.

Futrell, R.F., *et al.* 1961. *United States Air Force in Korea 1950-1953*. New York: Duell, Sloan & Pearce, 774 pp + pl.

Goldblat, J. 1982. *Agreements for Arms Control: a Critical Survey*. London: Taylor & Francis, 387 pp.

Goldblat, J. 1984. The Environmental Modification Convention of 1977: an analysis. In: Westing, A.H. (ed.). *Environmental Warfare: a Technical, Legal and Policy Appraisal*. London: Taylor & Francis, 107 pp: pp 53–64 (Chap. 5).

Hersh, S.M. 1972. Rainmaking is used as weapon by U.S. *New York Times* 121(41,799):1–2. 3 Jul 1972.

IAEA. 1983. *Nuclear Power Reactors in the World*. Vienna: *International Atomic Energy Agency* Reference Data Series No. 2, 48 pp.

Kerr, R.A. 1982. Cloud seeding: one success in 35 years. *Science* (Washington) 217:519–521.

Krass, A.S. 1984. The Environmental Modification Convention of 1977: the question of verification. In: Westing, A.H. (ed.). *Environmental Warfare: a Technical, Legal and Policy Appraisal*. London: Taylor & Francis, 107 pp: pp 65–81 (Chap. 6).

Lane, H.U. (ed.). 1984. *World Almanac & Book of Facts*. 107th edn. New York: Newspaper Enterprise Association, 928 pp.

Liller, W. 1964. Optical effects of the 1963 Project West Ford experiment. *Science* (Washington) 143:437–441.

Manchee, R.J., *et al.* 1981. *Bacillus anthracis* on Gruinard island. *Nature* (London) 294:254–255, 295:362, 296:598.

Manchee, R.J., *et al.* 1983. Decontamination of Bacillus anthracis on Gruinard island? *Nature* (London) 303:239–240.

Marshall, E. 1980. Iraqi nuclear program halted by bombing. *Science* (Washington) 210:507–508.

Marshall, E. 1981. Fallout from the raid on Iraq. *Science* (Washington) 213:116–117, 120.

Mason, J. 1980. Review of three long-term cloud-seeding experiments. *Meteorological Magazine* (London) 109:335–344.

Mészáros, E. 1984. Techniques for manipulating the atmosphere. In: Westing, A.H. (ed.). *Environmental Warfare: a Technical, Legal and Policy Appraisal.* London: Taylor & Francis, 107 pp: pp 13–23 (Chap. 2).

Moore, J. (ed.). 1984. *Jane's Fighting Ships, 1983-84.* 86th edn. London: Jane's Publishing Co., 779 pp.

Noltimier, H.C. 1984. Techniques for manipulating the geosphere. In: Westing, A.H. (ed.). *Environmental Warfare: a Technical, Legal and Policy Appraisal.* London: Taylor & Francis, 107 pp: pp 25–31 (Chap. 3).

Pimentel, D., *et al.* 1980. Environmental quality and natural biota. *BioScience* (Washington) 30:750–755.

Ramberg, B. 1980. *Destruction of Nuclear Energy Facilities in War: the Problem and the Implications.* Lexington, MA, USA.: Lexington Books, 195 pp.

Rees, D. 1964. *Korea: the Limited War.* New York: St. Martin's Press, 511 pp + 15 pl.

Stein, D.L. 1983. Electromagnetic pulse: the uncertain certainty. *Bulletin of the Atomic Scientists* (Chicago) 39(3):52–56.

Stevenson, A.E. 1963. U.S. replies to Soviet charges against certain space activities. *Department of State Bulletin* (Washington) 49:104–107.

Sullivan, W. 1975. Ozone depletion seen as a war tool. *New York Times* 124(42,769):20. 28 Feb 1975.

Sullivan, W. 1983. Scientists ponder forcing asteroids into safe orbits. *New York Times* 132(45,548):C3, C8. 4 Jan 1983.

Turco, R.P., *et al.* 1983. Nuclear winter: global consequences of multiple nuclear explosions. *Science* (Washington) 222:1283–1292.

UNGA. 1982. *World Charter for Nature.* New York: *United Nations General Assembly Resolution* No. 37/7 (28 Nov 82), 5 pp.

Westing, A.H. 1976. *Ecological Consequences of the Second Indochina War.* Stockholm: Almqvist & Wiksell, 119 pp + 8 pl.

Westing, A.H. 1977. Geophysical and environmental weapons. In: Westing, A.H. (ed.). *Weapons of Mass Destruction and the Environment.* London: Taylor & Francis, 95 pp: pp 49–63 (Chap. 3).

Westing, A.H. 1980. *Warfare in a Fragile World: Military Impact on the Human Environment.* London: Taylor & Francis, 249 pp.

Westing, A.H. 1981. Environmental impact of nuclear warfare. *Environmental Conservation* (Cambridge, UK) 8(4):269–273.

Westing, A.H. 1982. Environmental consequences of nuclear warfare. *Environmental Conservation* (Cambridge, UK) 9(4):269–272.

Westing, A.H. (ed.). 1984. *Herbicides in War: the Long-term Ecological and Human Consequences.* London: Taylor & Francis, 210 pp.

Westing, A.H. *et al.* 1984. Environmental warfare: policy recommendations. In: Westing, A.H. (ed.). *Environmental Warfare: a Technical, Legal and Policy Appraisal.* London: Taylor & Francis, 107 pp: pp 83–88 (Chap. 7).

Westman, W.E. 1977. How much are nature's services worth? *Science* (Washington) 197:960–964.

Wet, C.R. de. 1902. *Three Years War (October 1899–June 1902* (translated from the Dutch). Westminster, England: Archibald Constable, 520 pp + 1 map.

Chapter 7
Nuclear War: Its Environmental Impact

Note *: The use of nuclear weapons by the USA in 1945 was soon followed by their development and announced adoption by China, France, the United Kingdom, and the USSR. It was amply demonstrated in Japan that the further employment of those weapons even, in single or small numbers, could have devastating impacts on targeted humans and their associated infrastructures. Then, especially during the middle years of the East–West Cold War (which* in toto *lasted from ca 1946 to 1991), there existed a continuing widespread fear of a major nuclear war with its incredible levels of destruction. And that fear was hardly abated by non-binding declarations by four of those five great powers (not including the USA) that they would never be the first to use a nuclear weapon. Adoption by all nuclear powers of a no-first-use treaty could be a useful step in the right direction (#195).*[1] *Another interim step towards nuclear sanity would be for the nuclear powers to categorically renounce and destroy their tactical nuclear weapons, whether doing so individually or collectively.*

In time I felt it necessary to examine in some detail one aspect of nuclear war that in my view was being given inadequate attention, namely its potential for overwhelming environmental disruption. I thus began looking into the environmental effects of the use of nuclear weapons of various sorts, both individually and in time in the large numbers that might well have been expended in a possible major East–West nuclear holocaust. My reports began with a chapter in my Weapons of Mass Destruction and the Environment *(#97), this followed by a special look at the so-called neutron bombs (#102), that in turn followed by informed speculations on the impact of a limited nuclear war (#119, #131, #134, #140). During the Cold War period, sometimes in partnership with others, I also testified in both the US Senate and US House of Representatives, as well as producing a number of lesser journal and newspaper articles. Finally, what is reproduced below (****#176****) was my careful examination of the projected outcome of*

[1] The numbered references are provided in Chap. 3.

A. H. Westing, *Arthur H. Westing*, SpringerBriefs on Pioneers in Science and Practice 1, DOI: 10.1007/978-3-642-31322-6_7,

a major nuclear war, a paper, it was encouraging to learn, that was recognized as the best environmental paper of the year by the Foundation for Environmental Conservation.[2]

Abstract This paper examines the widespread environmental effects, sensu stricto, that would result from a large-scale nuclear war and the resultant ecological impacts. Singled out for analysis are the effects of wildfires, radioactive fallout, enhanced ultraviolet radiation, loss of atmospheric oxygen, gain in atmospheric carbon dioxide, and reductions in sunlight and temperature; also of combinations and ramifications of these adverse phenomena.

In a large-scale nuclear war, *wildfires* initiated by the heat of the nuclear explosions would burn on perhaps 2 % of the rural portions of the targeted countries' lands. These conflagrations would kill off much of the flora and fauna on those lands, and would cause long-term site debilitation through loss of soluble nutrients ('nutrient-dumping') and soil erosion. *Radioactive fallout* derived from surface bursts and destroyed nuclear reactors would for several days subject perhaps 10 % of the rural portions of the targeted countries to nuclear radiation (gamma, beta, etc.) at levels that would be lethal to all exposed vertebrates and to some categories of vegetation—for example, coniferous forests. At least 1 % of the territory of these countries would remain uninhabitable to humans for many decades. The killed vegetation would in time provide the fuel for further wildfires.

Ultraviolet radiation-B (UV-B) would be intensified perhaps threefold throughout the northern hemisphere (and to a lesser extent in the southern hemisphere) through depletion of stratospheric ozone by oxides of nitrogen generated by the fireballs created by atmospheric nuclear explosions, diminishing to normal levels only over a period of some years. This UV-enrichment would debilitate both the agricultural and natural ecosystems on land, and perhaps also those in the ocean, for some years. *Atmospheric oxygen* would be reduced through respiration that would be unbalanced by photosynthesis, and in other ways including conflagration, but at biologically and ecologically inconsequential levels. *Atmospheric carbon dioxide* would be enhanced through the same process of respiration unbalanced by photosynthesis, and in other ways including conflagration, by somewhat less than 20 % (perhaps quickly adjusting to somewhat less than 10 %). This CO_2-enrichment would have only minor ecological effects, but would presumably contribute to the coming global 'greenhouse' problem.

---

[2] Reproduced from: *Environmental Conservation* (Cambridge, UK) 14(4):295–306; Winter 1987 with the original title: *"The Ecological Dimension of Nuclear War"*, by permission of the Foundation for Environmental Conservation, the copyright holder, on 26 March 2012. Based on an invited paper given at the International Conference on Protection of the Environment and the Defence of World Peace, Varna, 25–29 August 1986, of the Bulgarian State Committee for Protection of Nature. The author is most pleased to acknowledge help in the final realization of this paper from Lawrence C. Bliss (University of Washington), J. Graham Cogley (Trent University), Lynn Miller (Hampshire College), Peter A. Schmidt (Technische Universität Dresden), and John M. Teal (Woods Hole Oceanographic Institution).

Sunlight would be partially obscured by smoke from urban and industrial fires, etc., in great moving patches throughout the mid-latitudes of the northern hemisphere for up to a few weeks. Although such diminution of light would have little effect per se on the biotas, it might do so secondarily by causing more or less drastic reductions in temperature. *Surface temperatures* might sporadically drop conceivably by as much as several tens of degrees Celsius in the interior of continental land masses of the mid-latitudes of the northern hemisphere for up to a few months, whereafter modest temperature depressions might continue for a year or so. Thus, if a large-scale nuclear war were fought especially during the growing (summer) season, the reduced temperatures would kill or injure crops, livestock, and the natural flora and fauna. Among the natural ecosystems, those of deciduous (broad-leafed) forests would be most seriously affected, with ecological recovery taking a considerable number of years. In the considerably less likely event that the cold wave would extend into the tropical (hitherto frost-free) regions, ecological damage would be catastrophic.

The various predicted environmental perturbations are of such magnitude in areal extent, intensity, and diversity that they would be sure to produce a variety of unforeseen ecological effects, especially in their interactions. The human survivors of a large-scale nuclear war would have to cope with a bleak and widely inhospitable environment indeed. It is concluded that nuclear war must be avoided, not only as the ultimate insult to human civilization, but also as the ultimate insult to nature.

7.1 Introduction

The immediate human impact of a large-scale nuclear war would be appalling. The human fatalities plus serious injuries and illnesses from such a cataclysm could add up to a substantial fraction of our global population. The societal infrastructures of the major nations would presumably be obliterated as well in such a war—including many of the large cities of those nations, their national and international communication and transportation systems, most of their medical facilities, their ability to pursue mechanized agriculture, their administrative and governance hierarchies, and, of course, such other amenities of life as their educational systems and the artifacts that comprise their cultural heritages (Bergström *et al.*, 1987; Chazov *et al.*, 1984; Daugherty *et al.*, 1985–1986 ; Din & Diezi, 1984; Katz, 1982; Sharfman *et al.*, 1979; Solomon & Marston, 1986; Thunborg *et al.*, 1981).

My present analysis, however, does not dwell upon the terrible human and social impacts of the hypothetical large-scale nuclear war just alluded to, but rather upon its foreseeable widespread, longer-term environmental effects and the resultant ecological changes. These environmental disruptions and their ecological ramifications are tragic—both for the effect they would have upon the immediate survivors of a nuclear holocaust, and in their own right.

The ecological dimension of nuclear war is discussed under the following six headings: (a) large-scale wildfires (briefly); (b) radioactive fallout (also briefly);

(c) enhancement of ultraviolet radiation; (d) loss of atmospheric oxygen and gain in carbon dioxide; (e) reduction of sunlight and temperatures (in some detail); and (f) overall effects.

The horrendous immediate impact of nuclear detonations on the local ecosystems, both terrestrial and marine, is not covered here (for this cf., *e.g.,* Westing, 1977, pp 2–30; 1980, pp 154–163). This analysis benefits from a number of past treatments of the subject (*ACDA*, 1978; Bensen & Sparrow, 1971; Ehrlich, P.R., *et al.*, 1983; Glasstone & Dolan, 1977; Greene *et al.*, 1985; Harwell, 1984 [cf. Westing, 1985b]; Larsson, 1981; Nier *et al.*, 1975; Pittock *et al.*, 1985–1986 [cf. Westing, 1986]; Rotblat, 1981; Schultz & Whicker, 1985; Svirezhev *et al.*, 1985; Westing, 1977; 1978; 1981; 1982).

7.2 Large-Scale Wildfires

A nuclear bomb dissipates roughly one-third of its explosive energy in the form of an intense thermal or heat wave. This heat would initiate wildfires over an immense area, of which the exact size would depend, of course, upon the weather conditions at the time, the terrain, and the nature of the vegetative cover. Indeed, under certain terrain and fuel conditions these fires would coalesce into a truly infernal firestorm. On a clear, dry summer day a single 1 megaton (MT) air burst might well initiate wildfires throughout an area of more than 33,000 hectares, and these could continue to burn and spread for days (Westing, 1977, pp 7–10).[3]

These fires would create havoc among the plants and animals that survived the blast and nuclear radiation. The surface disruption from blast and fire would in turn lead to massive site degradation of long duration (*i.e.,* decades) from nutrient losses in solution (so-called 'nutrient dumping') and soil erosion. The fires would also inject immense amounts of smoke into the atmosphere, the effects of which are treated in a separate section below.

In the event of a large-scale nuclear war, one could speculate with respect to the USA, for example, that as many as 1,000 distinct (non-overlapping) rural sites might conceivably be subjected to nuclear attack. With the simplifying assumption that each of these sites receives and detonates a single 1 MT bomb or its equivalent, then especially during the summer months some 30 million hectares would be subject to immediate ignition. A modest amount of autonomous (self-propagating) spread of fires could also be expected at many locations, increasing the overall fire area by several percent. Perhaps 15 % of the rural sites would be forestland, 25 % grassland, and the remaining 60 % cropland (leaning, for these proportions, on the summary of Small & Bush, 1985). Severe fire damage could

---

[3] A 1 megaton (MT) bomb has an energy yield equivalent to 907×10^6 kg of trinitrotoluene (TNT), *i.e.,* a yield of 4.19×10^{15} J. A 1 kiloton (kt) bomb has a yield one-thousandth of this value. The characteristics of nuclear bombs are summarized elsewhere (Westing, 1980, pp 180–181).

thus occur to as much as ca 2 % of the area of the conterminous USA, and of the order of 3 % of its forest, grassland, and lesser biomes (ecosystems). (These areas of fire damage are comparable with the upper end of the range of speculations of Small & Bush [1985], and with the lower end of those of Crutzen *et al.* [1984] or of Svirezhev *et al.* [1985, pp 70–73].)

To generalize, it can be concluded as a rough approximation that, in a large-scale nuclear war, wildfires would burn on perhaps 2 % of the rural portions of the various targeted countries.

7.3 Radioactive Fallout

A nuclear bomb dissipates about one-tenth of its explosive energy in the form of nuclear radiation, a portion of which is released as an initial burst, but the remainder—in the form of radioactive fallout—only much more slowly and widely. A single 1 MT ground burst would present a lethal dosage of nuclear radiation (gamma, beta, etc.) to all exposed vertebrates including, of course, humans and livestock (*i.e.,* > 2 kR), over about 36,000 hectares. As a highly instructive example I might cite the 'Bravo' test at Bikini on 1 March 1954—a single 15 MT ground burst—which, during the first 4 days, is known to have resulted in fallout that would be lethal to exposed humans and livestock (*i.e.,* > 0.5 kR) over an area of almost 2 million hectares, that is, over an area approximately half the size of Belgium (Westing, 1977, p. 14). Moreover, although the test programme at Bikini ended more than a quarter of a century ago, the atoll remains uninhabitable despite intensive cleanup attempts (Alcalay, 1980; Johnson, 1980).

It must also be noted here that, in a war, some of the hundreds of stationary or mobile (naval) nuclear reactors and other facilities containing large quantities of radioactive elements would be hit, and that the thereby-dispersed radioactive debris would increase the extent, in both area and time, of the regions lethal to flora and fauna, and uninhabitable by humans (Behar *et al.*, 1987; Fetter & Tsipis, 1981; Flavin, 1987; Hippel & Cochran, 1986; Hohenemser *et al.*, 1986; Petersen *et al.*, 1986; Ramberg, 1984).

In the event of a large-scale nuclear war, one could once again, by way of example, speculate with respect to the USA that there might conceivably be as many as 2,000 nuclear bombs or warheads (here again assumed to average 1 MT each) detonated as surface bursts in as many distinct (non-overlapping) locations. With the assumption that each resulted in a separate 36,000-hectare zone of fallout that would be fatal to all exposed vertebrates—including, of course, all exposed humans and livestock—and to all conifer systems as well (> 2 kR), this would represent ca 9 % of the area of the conterminous USA and of the order of 13 % of its forest, grassland, and some lesser biomes. (Although not directly comparable, these areas are compatible with those of Pittock *et al.* [1985–1986, I, pp 249–250], especially in view of the huge uncertainties involved.)

To generalize again, it can be concluded as a rough approximation that, in a large-scale nuclear war, radioactive fallout would, for several days, subject perhaps 10 % of the rural portions of the targeted countries to nuclear radiation at levels that would be lethal to all exposed vertebrates and conifers. At least 1 % of the territory of those countries would remain uninhabitable to humans and unusable for agriculture for many decades.

7.4 Enhancement of Ultraviolet Radiation

It has been estimated that the many detonations in a large-scale nuclear war in the northern hemisphere (a region that accounts for 75 % of the world's habitable land area and fully 90 % of its population) would, through their intense heat, generate sufficient oxides of nitrogen to eliminate about half (30 % to 70 %) of the stratospheric ozone throughout that hemisphere (and some lesser fraction in the southern hemisphere), such areas not returning to normal for a period of perhaps several years (Carrier *et al.*, 1985; Nier *et al.*, 1975; cf. also Holdsworth, 1986).

Such a reduction in the stratospheric ozone layer would in turn lead to an approximately threefold increase in the amount of ultraviolet radiation that reaches the earth's surface in the biologically active range of 280–315 nm, the so-called UV-B (Gerstl *et al.*, 1981). The ecological impact of enhanced UV-B radiation on various natural and artificial ecosystems—oceanic, terrestrial, and agricultural—cannot be predicted with any certainty, but might possibly be devastating on a world-wide basis (*ACDA*, 1978 ; Kruger *et al.*, 1982; Nier *et al.*, 1975; Tukey *et al.*, 1979).

Oceanic ecosystems might possibly be globally disturbed by the enhanced UV-B radiation following a large-scale nuclear exchange (Calkins & Thordardottir, 1980; Jokiel, 1980; Kruger *et al.*, 1982; Tukey *et al.*, 1979; Worrest *et al.*, 1981a ;1981b; Worrest, 1983). Some portion of the marine plankton lies close to the surface of the ocean, and if it turned out that this fraction were killed off to a substantial extent, the oceanic food chain would be in part disrupted and the fish stocks at the upper end of this chain thereby placed in jeopardy through starvation. As concomitant ozone depletion is assumed to become more or less global in extent, repopulation by the phytoplankton might take a number of years, and would then be unlikely to be in time to save some more or less small fraction of the many plankton-dependent species. Restoration of the depleted fish stocks throughout the world (both commercial and otherwise) might thus in turn take many years to occur.

Terrestrial ecosystems might possibly also be substantially disrupted on a global basis by the enhanced UV-B radiation following a large-scale nuclear exchange (Caldwell, 1981; Faber *et al.*, 1979; Kruger *et al.*, 1982; Nier *et al.*, 1975; Teramura, 1983; Tukey *et al.*, 1979). This is so because it can be extrapolated from the (admittedly very limited) available information, that possibly as many as 20 % of the plant species might succumb directly or indirectly, and an additional fraction of them would have their photosynthesis and food production, as well as growth, impaired. And these debilitations would, of course, add to those more regional ones of nuclear radiation from fallout (especially so in the case of the relatively sensitive conifer ecosystems), and so forth.

Such possibly drastic perturbations among some of the primary producers of the world's ecosystems (perturbations that would include newly-altered relationships of competitive advantage among species) could in turn be expected to exert a substantial impact on the dependent animal wildlife throughout the world. Some of the animal life might conceivably also be injured directly by the enhanced UV-B radiation. This is so because in most instances the postulated, newly-established damaging levels of UV-B would not be detectable by the animals, thereby precluding evasive actions. Thus, for example, damage to the cornea of the eye might reduce the efficiency of hawks, eagles, and much other wildlife, in their hunting or foraging abilities.

During the first several years following a major nuclear exchange, agricultural ecosystems would possibly also be severely disrupted on a more or less global basis by the enhanced UV-B radiation, although relevant data are quite limited (Teramura, 1983). And again, such damage would presumably compound the problems resulting from regional radioactive contamination (to which crops are generally more sensitive than their weed competitors and their fungal and insect pests) and from a paucity of farm workers, implements, fuel, fertilizers, and pesticides (herbicides, fungicides, insecticides). Some crops—among them sugar beets, tomatoes, beans, peas, and perhaps corn (maize) and rice—appear to be especially sensitive to enhanced UV-B radiation; and livestock might develop debilitating corneal lesions and perhaps also skin lesions.

A final word of caution is in order regarding these predictions of UV-B-engendered damage, based as they are on such limited data. As a possible counter-example, the Tunguska meteor fall of 30 June 1908 may have resulted in a world-wide depletion of stratospheric ozone—and thus of UV-B enrichment—of several years' duration, comparable with that following a rather large-scale nuclear war (Park, 1978; Turco *et al.*, 1981; 1982; but cf. Rasmussen *et al.*, 1984). If the suggested UV-B enrichment did indeed occur, then it seems not to have had an effect on the exposed biota. To begin with, an analysis of the relevant weather records by Turco *et al.* (1982, pp 42–43) could establish no significant post-meteor anomalies.

An examination by me of the relevant production and land-use data tabulated by Mitchell (1981) for wheat, oats, potatoes, and sugar beets—in Sweden, France, and Italy—provides no indication on either a total or unit-area basis of post-meteor adversities. Moreover, there appears to be no contemporary or other historical account of the period that alludes to ecological or other biotic calamities which might have been the result of enhanced UV-B radiation.

7.5 Loss of Atmospheric Oxygen and Gain in Carbon Dioxide

7.5.1 Basic Data

It is suggested from time to time that a large-scale nuclear war would deplete the earth's atmospheric diatomic oxygen (O_2) by a dangerous if not catastrophic amount (*e.g.*, Gromyko *et al.*, 1984, pp 19–20). I contend that an oxygen depletion of such magnitude is not possible. In my dismissive argument I find it convenient to discuss also the question of atmospheric carbon dioxide (CO_2) enhancement.

The atmosphere can be calculated to have a total mass of ca 5.117×10^{18} kg. Its oxygen component weighs ca 1.186×10^{18} kg; and there has been no discernible change in this value in modem times. Its carbon dioxide component now weighs ca 2.662×10^{15} kg, which represents an increase of about 21 % from the ce 1800 value of ca 2.195×10^{15} kg, owing to anthropogenic additions.[4]

The annual addition of oxygen to the atmosphere owing to photosynthesis by the autotrophs (green plants) of the world is ca 208×10^{12} kg; and the annual subtraction owing to respiration by all living organisms (autotrophs plus heterotrophs, *i.e.*, vertebrates, invertebrates, bacteria, fungi, etc.) is assumed to be essentially the same. The annual subtraction of carbon dioxide from the atmosphere owing to photosynthesis is ca 286×10^{12} kg; and the annual addition owing to respiration is assumed to be essentially the same. Photosynthesis and respiration appear to be the only significant natural sources of atmospheric gain or loss, respectively, for these two molecules.[5]

Photosynthesis, and presumably also respiration, are distributed throughout the globe as follows (Westing, 1980, pp 20–22): terrestrial northern hemisphere, 51 %; terrestrial southern hemisphere, 18 %; oceanic northern hemisphere, 13 %; and oceanic southern hemisphere, 18 %. Moreover, the vegetated land area of the

---

[4] The mass of the air (density, 1.292 kg/m^3) is based upon a global surface area of 510.1×10^{12} m^2, of which 361.3×10^{12} m^2 (70.8 %) is ocean and 148.8×10^{12} m^2 (29.2 %) is land, the latter having an average elevation, with ice caps, of 780 meters above mean sea level. The air pressure at sea level is 10,330 kg/m^2, and that at 780 meters is 9,230 kg/m^2 (and thus that at 228 meters, the average global elevation, is 10,031 kg/m^2). The average land elevation, with ice caps, of 780 meters was calculated by J.G. Cogley (Trent University, Peterborough, Ontario, private communication, 4 November 1985; based on Cogley, 1985, Tbl A1). The mass of the oxygen (density, 1.429 kg/m^3) is based upon a concentration in air of 209.5 L/m^3. This concentration has remained constant for millennia (Broecker, 1970; Holland, 1978, pp 284–295). The masses of the carbon dioxide (density, 1.977 kg/m^3) are based upon a ce 1800 concentration of 280 mL/m^3 (Neftel *et al.*, 1985; Pearman *et al.*, 1986) and a present concentration of 340 mL/m^3 (Keeling *et al.*, 1982).

[5] The photosynthetic gain/loss values for oxygen and carbon dioxide are based upon a global annual net primary production of ca 171.5×10^{12} kg, with a carbon content of 45.5 % (Westing, 1980, pp 21–22); an atomic weight of carbon of 12; and a mole-for-mole equivalence of oxygen (molecular weight, 32) and carbon dioxide (molecular weight, 44) owing to their mole-for-mole equivalence in the photosynthetic equation. Net changes in atmospheric oxygen and carbon dioxide as a result of weathering, sedimentation, volcanic action, and other natural physical processes, are apparently trivial in magnitude (Holland, 1978, pp 270–295).

globe covers ca 133 × 10^{12} m^2, of which 99 × 10^{12} m^2 (74 %) is in the northern hemisphere and 34 × 10^{12} m^2 (26 %) is in the southern hemisphere (Westing, 1980, p. 20). The estimated total global terrestrial dry weight of plant biomass (both above- and below-ground) is ca 1.84 × 10^{15} kg, giving an average of 13.8 kg/m^2 of vegetated land (Westing, 1980, pp 21-22). Perhaps 60 % of terrestrial plant biomass is above ground, giving a global average of 8.28 kg/m^2.

7.5.2 Environmental Effects

In the event of a large-scale nuclear war, depletion of atmospheric oxygen and enhancement of atmospheric carbon dioxide would be brought about in two principal ways: (a) through the respiration of surviving heterotrophs that had not been replaced by the photosynthesis of autotrophs; and (b) through the action of wildfires. Moreover, as a third avenue of change, (c) both oxygen and carbon dioxide would be somewhat depleted through the action of the intense heat generated by the nuclear explosions. Each of these three sources of change is discussed below.

Global photosynthesis might be substantially reduced for several years or so owing to: (a) the reduced light and especially the reduced temperature that might be brought about by an atmospheric soot load; and (b) the enhanced UV-B radiation that might be brought about by reduced stratospheric ozone. Global photosynthesis might, for these reasons, be reduced—at least for purposes of the present argument—by 75 % during the first post-war year (assuming reductions in the terrestrial northern hemisphere by ca 100 %, in the terrestrial southern hemisphere by ca 50 %, and in the global ocean by ca 50 %). This would be followed by gradual recovery: the global reduction in photosynthesis during the second post-war year might be 50 %, during the third such year 25 %, during the fourth year 12 %, and during the fifth year 5 %. Heterotrophs would also be globally depleted, but by a relatively smaller amount, and similarly experiencing gradual recovery. The global reduction in heterotrophs might be 10 % in the first post-war year and 5 % in the second.

Given the above order-of-magnitude assumptions on reductions in global photosynthesis and respiration, one can calculate that the loss of oxygen to the atmosphere from respiration not balanced by photosynthesis would be ca 135 × 10^{12} kg during the first post-war year and ca 318 × 10^{12} kg during the first five post-war years. Thus, losses in the world's biotas could lead to a five-years' decrease in atmospheric oxygen of 0.03 %. The simultaneous gain to the atmosphere in carbon dioxide owing to the excess respiration, would be ca 186 × 10^{12} kg during the first post-war year, and ca 438 × 10^{12} kg during the first five post-war years. Thus, losses from the biotas could lead to a five-year increase in atmospheric carbon dioxide of 17 %.

Wildfires (for purposes of argument) might conceivably rage over as much as 3 % of the vegetated land-area of the northern hemisphere, that is, over as much as ca 2.97×10^{12} m^2. The above-ground dry-weight biomass on this, perhaps exaggerated, area is ca 24.6×10^{12} kg, containing ca 11.2×10^{12} kg of carbon. Assuming that a fire consumes as much as one-third of the above-ground biomass (containing 3.7×10^{12} kg of carbon), then the amount of oxygen subtracted from the atmosphere would be ca 9.9×10^{12} kg, and the amount of carbon dioxide added would be ca 13.7×10^{12} kg. (These values are based on the atomic weight of carbon [12] and the molecular weights of oxygen [32] and carbon dioxide [44].) Thus, wildfires could lead to a decrease in atmospheric oxygen of 0.001 %; and to an increase in atmospheric carbon dioxide of 0.5 %.

Nuclear explosions (perhaps as many as 16,000) could conceivably add up to a combined yield of 10,000 MT. At an assumed high-temperature conversion, in the fireballs, of oxygen molecules to other chemical 'species' (*e.g.*, NO_2) at the rate of ca 4×10^6 kg/MT (Westing, 1977, pp 9–10), this would come to a loss of 40×10^9 kg of atmospheric oxygen. At a speculated high-temperature conversion in the fireballs of carbon dioxide to other chemical 'species' at the molecule-for-molecule proportional rate of ca 6×10^3 kg/MT, this would come to a loss of 60×10^6 kg of atmospheric carbon dioxide. Thus, nuclear explosions could lead to a decrease in atmospheric oxygen of 0.000003 %, and to a decrease in atmospheric carbon dioxide of 0.000002 %.

7.5.3 Ecological Consequences

An atmospheric *oxygen* diminution of 0.031 % (0.03 % + 0.001 %) is biologically and ecologically inconsequential. To provide some sort of perspective, oxygen per unit volume diminishes by fully 1 % in going from sea level to an elevation of only 78 meters. In Denver, Colorado, at an elevation of 1,609 m, there is 19 % less oxygen per unit volume than at sea level.

An atmospheric *carbon dioxide* enhancement of nearly 18 % (17 % + 0.5 %) might have a slight ecological effect. As noted earlier, human actions have already enriched the carbon dioxide content of the atmosphere by about 21 % over the past 185 years or so. A probable global result of this additional global increase would presumably be a slight enhancement of the 'greenhouse' effect of atmospheric carbon dioxide, that is, a slight warming of the surface of the earth. There might also be a slight global enhancement of photosynthesis inasmuch as the atmospheric carbon dioxide level might, for some plants, be the limiting factor in this process.

A carbon dioxide increase of the magnitude assumed would be inconsequential in terms of biological toxicity with respect to humans or other animals, plants, or micro-organisms. Actually, only about one-half of all past anthropogenic additions of carbon dioxide to the atmosphere have remained there (Clark *et al.*, 1982, p. 13 ; Watts, 1982, pp 458–460), so that some fraction of that inserted by a nuclear war might also soon leave the atmosphere.

7.6 Reduction of Sunlight and Temperature

7.6.1 *Background*

Several recent model calculations suggest that a large-scale nuclear war might inject into the atmosphere amounts of solid aerosols (fine dust and especially soot) sufficient to reduce substantially the incident solar radiation reaching the earth's surface—and thereby also reduce surface temperatures—for a period of some months (Aleksandrov & Stenchikov, 1983; Carrier *et al.*, 1985; Cess *et al.*, 1985; Cotton, 1985; Covey *et al.*, 1984; Crutzen *et al.*, 1984; Thompson, 1985; Turco *et al.*, 1983; Velikhov, 1985; cf. also Golding *et al.*, 1986; Malone *et al.*, 1986; Schneider, 1987; Westing 1985a). This effect could conceivably continue or prolong itself in one way or another (Robock, 1984; Warren & Wiscombe, 1985).

One prominent (much-quoted) hypothetical scenario (especially regarding ecological impact) is based upon a supposition of 16,160 nuclear explosions in the northern hemisphere, ranging between 100 kT and 10 MT, and totalling 10,000 MT, of which 6,300 MT of the total are surface bursts and of which 1,500 MT of the total are employed against urban or industrial targets (Turco *et al.*, 1983). One effect of such a 'maximum' or 'full' nuclear exchange was calculated to be a reduction in sunlight reaching the surface of the northern hemisphere that, on average, might attain a maximum of ca 98 %. Such a reduction to ca 2 % of full sunlight could be attained within about 1 week (*i.e.,* a reduction in the model from ca 170 W/m^2, the global average, to ca 3.3 W/m^2), could recover to 5 % of full sunlight in about 2 weeks from the beginning, to 10 % in about 3 weeks, and to 25 % in about 5 weeks, before returning to normal in about 40 weeks (estimated from Turco *et al.*, 1983, Fig. 4).

Such a reduction in incident light was in turn calculated to depress the ambient surface temperatures of the (otherwise undefined) interior land areas of the northern hemisphere by a maximum of 40 °C within 4 weeks (*i.e.,* a reduction in the model from 13 °C, the approximate global average, to −27° C), with a 10° C recovery (warming) occurring in about 8 weeks from the beginning, a 20° C recovery in about 14 weeks, a 30° C recovery in about 36 weeks, but without the time taken to return to normal being presented (estimated from Turco *et al.*, 1983, Fig. 2).

As an example, if a reduction in temperature of such magnitude and duration were initiated just before or during a growing season at a northern mid-latitude (*e.g.,* at the latitudinal level of Chicago, 42 °N, or Moscow, 56 °N), temperatures would remain depressed below freezing for the duration of that growing season; if, on the other hand, it were initiated early in a dormant season, the subsequent growing season would be shorter and would experience reduced (but not freezing) temperatures.

The drastic ecological impact of a light-plus-temperature perturbation of the environment of the sort postulated for a large-scale nuclear war, would appear to be confined to the inland continental land areas especially of the northern hemisphere and, in the temperate regions, would further seem to derive largely from the reduction in temperature that would occur during the growing season. If the temperature effects were to spread further southward, into northern-hemisphere areas that do not normally experience freezing temperatures (as they might do, according to Stenchikov [1985] and to Thompson [1985]), then the temperature depression could exert an impact at any time of the year.

There is, of course, no doubt that particulate aerosol matter in the atmosphere prevents some fraction of the sunlight from reaching the ground, and that this in turn results in reduced surface temperatures. Such cooling has been described as a consequence of both dust storms (Brinkman & McGregor, 1983) and of smoke from forest fires (Peterson & Drury, 1967; Wexler, 1950). The dust and sulfur aerosol injected into the atmosphere by volcanic action exerts a similar effect (Sear *et al.*, 1987).

The most pronounced example, from modem times, of aerosol injection into the atmosphere by a volcano, appears to be the eruption of Mount Tambora (8 °S 118 °E) on 10–11 April 1815 (Stothers, 1984). A number of regions far removed from the source of the aerosol experienced one or more serious cold spells during the summer of 1816. Prominent among the areas thus afflicted were portions of New England (especially Vermont) and also of western Europe; in those areas the inhabitants suffered serious crop failures (Hoyt, 1958; Post, 1974; Stommel & Stommel, 1979).

The eruption of Mount Tambora in 1815, the greatest in recorded history, was very spotty in its subsequent climatic impact, with severe temperature depressions during the following summer occurring in no more than a few regions of the world, and those only from time to time. The overall calculated mean temperature depression for the northern hemisphere in 1816 appears to have been only 0.4 °C (Stothers, 1984, p. 1197). Even though England, for example, is recorded as having had agricultural shortfalls in 1816 (Post, 1974), these must have been the result of a small number of cold snaps during the growing season, as my examination of the relevant temperature record at least for central England (Manley, 1974) reveals no anomaly. Moreover, even though western Europe was one of the afflicted regions, an examination by me of the relevant agricultural data tabulated by Mitchell (1981) for France, reveals no effect on the 1816 outputs of that country's wheat, rye, barley, oats, or potatoes, whether analyzed on a total or unit-area basis. Moreover, the dates of wine harvest (a sensitive indicator of weather change) were not demonstrably influenced in France during this period (Benarie, 1986). Finally it must be mentioned that it is not entirely clear that the noted reductions in temperature were, in fact, the direct result of the aerosol injected into the atmosphere by the volcano (Catchpole & Faurer, 1983, pp 135–136; Thompson & Schneider, 1985–1986, pp 176–177).

Thus it must be emphasized that the intensity, latitudinal and longitudinal spread, degree of patchiness, and duration of reduced light and temperature that might be brought about by a large-scale nuclear war, are all open to considerable question. The major limitations, uncertainties, and other inadequacies of the recent predictions, derive *inter alia* from an inability to foretell the size and nature of a future nuclear war, from an insufficient knowledge of meteorological physics and chemistry, and from being dependent upon computer models that are too simple to be able to cope with the problem (Barton & Paltridge, 1984; Covey *et al.*, 1985; Ehrlich, R., 1986; GAO, 1986; Golding *et al.*, 1986; Hecht *et al.*, 1985; Hobbs, 1985; Malone *et al.*, 1986; Paltridge & Hunt, 1984; Penner, 1986; Rathjens & Siegel, 1984–1985; Singer, 1984; 1985; Smith, 1985; Teller, 1984; Thompson & Schneider, 1985–1986; Weinberger, 1985–1986).

It is necessary to note in regard to the uncertainties associated with the effects of smoke, that a huge fire in the tropical forests of Borneo in 1983—one that devastated some 3.5 million hectares, or 5 % of that very large island—obscured the sun for weeks at a time, apparently without bringing about any low-temperature damage to either the flora or fauna that were at risk (MacDonald, 1984; Malingreau *et al.*, 1985). An even more immense concatenation of fires in Siberia in the summer of 1915 destroyed an estimated 14 million hectares of forest, producing vast quantities of smoke, but with only trivial impact on regional crop production (Seitz, 1986; Shostakovitch, 1925).

The great uncertainties aside, I present below the postulated effects on plants and animals of reductions in light and temperature of the sort that have recently been predicted.

7.6.2 Reduced Light

A reduction in light intensity per se of the severity and duration that is being predicted would have only a modest and transitory ecological impact, whether terrestrial or oceanic. Thus with regard to terrestrial plants, many species—the so-called *tolerant plants* (or shade plants)—have their total light requirements met (*i.e.,* are light-saturated) at between perhaps 5 % and 10 % of full sunlight; at the other extreme many species—most of the so-called *intolerant plants* (or sun plants)—have their total (or virtually total) light requirements met at between perhaps 20 % and 25 % of full sunlight. Moreover, tolerant plants generally continue to grow (albeit at reduced rates) when maintained at ca 0.5 % to ca 1 % of full sunlight (the so-called compensation point), as do many intolerant plants when maintained at ca 1 % to ca 3 % of full sunlight. To illustrate these points: in one 9-month trial, Redwood (*Sequoia*) continued to grow at 0.6 % of full sunlight; Spruce (*Picea* sp.) at 1.1 %; Douglas-fir (*Pseudotsuga menziesii*) at 1.4 %, and various species of Pine (*Pinus* spp) at 1.8 % to 2.7 % (Bates & Roeser, 1928; similarly, cf. Shirley, 1929).

Some plants, however, require higher levels of minimal sunlight than those just suggested: one species of pine that is native to semi-arid southwestern USA (Nut pine, *Pinus edulis*) was found to need at least 6.3 % of the equivalent of full sunlight for growth on a sustained basis. In an interesting experiment with the seedlings of 10 herbaceous plants, it was found that the 4 woodland (tolerant) species tested could survive about 8 weeks of total darkness at ca 16 °C, and perhaps three times that long at ca 6 °C (Hutchinson, 1967). The 6 open-habitat (intolerant) species tested could survive about 5 weeks of total darkness at the higher temperature and, again, about three times that long at the lower temperature.

A sample consisting of 95 temperate-forest tree species (most of the commercially important ones in North America) suggests that at least the temperate woody perennials are rather uniformly distributed along the spectrum of tolerance: very tolerant, 16 %; tolerant, 18 %; of average tolerance, 25 %; intolerant, 19 %; and very intolerant, 22 % (Toumey & Korstian, 1947, pp 350–353).

The oceanic flora (largely algal phytoplankton) is likely to be little affected by the predicted reduction in light per se, there probably being no effect in winter and only a transitory one in summer (a conclusion supported by the information presented by Milne & McKay [1982]).

With respect to animals, the predicted low light intensity per se would, I assume, not be a direct threat to the life of most species, but might well disturb hormonal balances, reproductive activity, and behavioral patterns.

7.6.3 Reduced Temperatures

In considering the effects of any expected low temperatures, it is important to distinguish between the winter-experiencing (boreal plus temperate) and frost-free (tropical) zones; between growing (summer) and dormant (winter) seasons; between plants and animals; and, for the plants, among herbaceous annuals or biennials, herbaceous perennials, and woody perennials. Of the approximately 250 thousand known vascular plant species that are alive today, about 21 % are indigenous to the north-temperate and boreal regions of the world, 65 % to the tropics, and the remaining 14 % to the south-temperate regions (Raven, 1976, p. 156; Raven *et al.*, 1981, pp 634–636).

A period of freezing initiated during a growing season in the temperate zone, would presumably kill many plants, substantially knock back (destroy portions of, *i.e.*, produce dieback in) or otherwise injure others, and do little damage to some (cf., *e.g.*, Altman & Dittmer, 1972–1974, pp 811–812). The majority of annuals and biennials (which together represent ca 23 % of all temperate-region vascular plant species and a considerably higher proportion of food crops) would be killed—an event that might well occur prior to their production of seeds for the next year, although in many instances seeds would have been safely stored in the soil from prior years. Herbaceous perennials (ca 56 % of all temperate-region vascular plant species) would either be killed or else (more likely) be knocked back to ground level; and, in the latter event, be ready to sprout at the beginning of the next growing season. The woody perennials—the remaining ca 21 % of all temperate vascular plant species, but representing virtually all in terms of plant biomass—presumably would variously be killed, seriously knocked back, moderately knocked back, or scarcely injured.[6]

I am aware of only one, slightly relevant, set of observations. In the Wasatch Mountains of Utah, USA (ca 39 °N), during the vigorous early portion, or height, of the growing season (*i.e.*,, at the plants' most tender time) in May 1919, there was a sudden extraordinary two-day cold wave that caused the temperature to plummet by 35 °C, namely from 26 °C to −9 °C (Korstian, 1921). Among the herbaceous (non-woody) plants present, practically all of the annuals were killed and the perennials were knocked back to ground level, though the latter subsequently sprouted up again. Among the dicotyledonous woody perennials present (hardwood shrubs and trees) there was a range of damaging though non-lethal effects. In most instances, however, wholesale injury resulted to the succulent tissues and even to some of the previous season's growth. Species of Maple (*Acer*), Alder (*Alnus*), Mountain-ash [Rowan] (*Sorbus*), Elder (*Sambucus*), and Oak (*Quercus*), were among those seriously injured, whereas a Currant (*Ribes* sp.) was less injured, and the small evergreen shrub, Mountain lover (*Pachystima* sp.) remained uninjured. Among the conifers, a range of effects was also observed: the several species of Fir (*Abies*) present were most seriously injured, some individuals of one of the species, Subalpine fir (*A. lasiocarpa*) even being killed; Douglas-fir (*Pseudotsuga*) was injured less severely than the true firs; the species of Spruce (*Picea*) present suffered only modest injury; and the species of Pine (*Pinus*) present were only slightly injured or even remained uninjured.

An apparently modest amount of indirect damage to the flora can be expected from a drop in temperature as a result of the resultant partial loss of bees and other pollinating insects (flies, beetles, moths, etc.). A major loss of pollinating insects would be disastrous for perhaps 3 % of the temperate vascular flora. The non-flowering vascular plants (lycopods, horsetails, ferns, gymnosperms, etc.), representing about 6 % of the vascular flora, are not dependent upon insects for their reproduction. Of the flowering plants (the angiosperms, *i.e.*, dicotyledons plus monocotyledons), about 22 % reproduce asexually, 23 % are self-pollinated, 41 % are wind-pollinated, and the remaining 14 % are insect-pollinated. About 77 % of the vascular flora is perennial and the remaining 23 % either annual or biennial. It

[6] The proportionate distribution of life forms within the temperate vascular flora presented here is based on counts by the present author among the 16,274 such species of plants for which appropriate information was tabulated by Shetler & Skog (1978), representing an estimated 90 % of the relevant indigenous North American flora. By way of comparison, the comparable ca 2,600 species growing in the [former] two Germanys break down as follows: annuals plus biennials, 26 %; herbaceous perennials, 62 %; and woody perennials, 12 % (P.A. Schmidt, Arboretum, Technische Universität Dresden, pers. comm., 30 September 1984).

is the annual plus biennial fraction of the insect-pollinated plants (*i.e.,* 3 % of the vascular flora) that would suffer the most from a loss of the necessary insects.[7]

A similar though less pronounced range of effects would be expected during the growing-season not only elsewhere in the temperate regions, but also in the more northerly regions, for at least some boreal plants appear to be nearly as cold-sensitive during the growing-season as temperate plants (Sakai & Otsuka, 1970). The postulated amount of reduction in temperature during the dormant season in either temperate or, especially, arctic regions could be expected to have minimal long-term effects on most flora (Parker, 1963; Sakai & Weiser, 1973), even more so because dormant-season temperature reductions are expected to be less severe than those in the growing-season (Malone *et al.*, 1986). The duration of freezing weather in those zones would probably also be of only minor importance per se, although extending a cold period does lead to some added damage (Parker, 1963). Moreover, the subsequent growing season would be cooler and shorter and thus very injurious to some species.

In the more southerly, frost-free regions of the northern hemisphere, and beyond, a period of freezing weather at any time of the year might well lead to wholesale plant damage and a rather high level of mortality—and thus conceivably also to a greater or lesser number of extinctions among those many species which are confined to the affected areas. Even near-freezing temperatures would cause much injury to tropical plants (Lyons, 1973).

Special emphasis should be given here to the effect of a shortened growing season with reduced, though non-freezing, temperatures. Such a condition might be expected in the southern fringe and coastal regions of the northern hemisphere, more widely in the southern hemisphere, and in the northern hemisphere during the growing season following a freezing one; or else as a result of a less than large-scale nuclear war, or if the changes were less drastic than has recently been suggested by some as likely. It has been quite well documented in quantitative terms that such a perturbation could have an adverse effect on the growth of plants—including, of course, crop plants (Westing, 1977, p. 28). Thus, for Soybeans (*Glycine max*) in central USA, each 1 °C decrease from the optimal temperature during one month in the middle of the growing season is apt to decrease the season's yield by 2.5 %. Moreover, a seemingly very modest cooling or, perhaps more importantly, shortening of the growing season, will spell the difference between success or failure of a crop that is being grown near the limit of its

[7] Estimates of the number of species within each division of vascular plants are provided by Jones (1951) and by Raven *et al.* (1981, pp 634–636). The proportionate distribution of modes of reproduction within the vascular flora presented here is based on counts by the present author from among the 1,488 angiosperm species (within 128 families) for which appropriate information was tabulated by Fryxell (1957), representing somewhat more than 0.6 % of the world-wide total (or perhaps 3 % of the relevant temperate flora, from which most of the sample was drawn). The proportionate distribution of wind *versus* insect pollination among the cross-fertilized species is a mid-temperate-zone estimate by the present author, based on the compilation and analysis by Regal (1982).

range (cf., *e.g.*, Bergthorsson, 1985; Hare *et al.*, 1985, pp 37–41; Pittock *et al.*, 1985–1986, II, pp 304–312). Conversely, under some conditions—for example, under the water-limited conditions of grasslands—modest reductions in light and temperature might actually prove to be beneficial to the system (McNaughton *et al.*, 1986).

Turning now to the fauna, the effect of a period of freezing summer weather is difficult to predict with any confidence. Much animal wildlife, especially the herbivores, would find it hard to obtain sufficient food. The warm-blooded animals in general (all of the mammals and birds) would presumably suffer a higher level of mortality than the cold-blooded ones (the reptiles, amphibians, fishes, and insects and other arthropods). Many of the warm-blooded animals would survive exceedingly low temperatures providing they found enough to eat (Hensel *et al.*, 1973, pp 659–660 & 666–667; cf. also Altman & Dittmer, 1972–1974, pp 784–793). Many of the cold-blooded animals would presumably survive by going dormant or into a state of cold torpor until their habitats became warm again (Laudien, 1973, pp 441–443; Precht, 1973, pp 410–419; cf. also Altman & Dittmer, 1972–1974, pp 794–807; Lee *et al.*, 1987).

Life in the ocean would presumably not be seriously affected by the predicted drop in air temperature, because of the thermal inertia of water and the sufficiently wide tolerance of the marine biota. However, some mortality would doubtless result if a serious drop in the water temperature were, in fact, to occur. A transitory temperature drop of several degrees Celsius in the Persian Gulf in 1964 (an extraordinary event) led to extensive mortality among the indigenous fishes, sea snakes, and cuttlefish, and especially the corals (Shinn, 1976, pp 251–252; cf. also Gunter & Hildebrand, 1951).

7.7 Overall Effects

Beyond the hideous immediate and relatively close-in effects on the biotas of blast, heat, and ionizing radiation, from each of the many thousands of nuclear bombs that one presumes would be expended in any future nuclear war, surviving plants and humans and other animals would be exposed to radioactive fallout throughout huge geographical areas. Moreover, the radioactive contamination would be exacerbated, both as to area and duration, to the extent that the hundreds of existing nuclear facilities had been targeted and their contents scattered. In some areas, the wildfires initiated would continue to burn and spread for many days. In time, large areas of radiation-killed forest (especially conifer forest) would provide the fuel for further extensive wildfires. All of the burnt-over areas would be additionally debilitated by 'nutrient dumping' (loss of nutrients in solution) and soil erosion.

To the local and regional effects must be added the possibility of ultraviolet (UV-B) radiation at levels that would be injurious to the biotas (both terrestrial and marine)—an impact which would presumably be more or less global in extent and of several years' duration. And the most recent specter raised has been the possibility of up to several months of very cold weather throughout the interiors of the land masses of the northern hemisphere, if not of the whole world. An effect of this sort would disrupt agriculture and would kill much wildlife and vegetation. The ecological impact of such reductions in temperature would be considerably less pronounced if the war occurred during the dormant (winter) season, and more so if it were during the growing (summer) season. The impact would be less severe at the higher (Pole-ward) latitudes and more so at the lower (Equator-ward) latitudes. Indeed, if the tropics were subjected to such reductions in temperature, the results could well be calamitous in terms of plant and animal deaths and even species extinctions.

It will be useful to suggest the impact of the predicted reductions in temperature separately for each of the world's several major biogeographical zones. Of the 133×10^{12} m^2 of land areas on earth which support communities of plant and animal life (Westing, 1980, pp 20–23): ca 12 % have been converted to agriculture (taken largely from both the temperate and tropical zones); ca 6 % support arctic or alpine tundra biomes (those regions having too short a growing season, or being otherwise too climatically harsh, to permit the development of trees or expanses of grasses); ca 19 % support desert (arid) biomes (those regions, whether temperate or tropical, that are too dry to permit the development of forests or grasslands); ca 15 % support temperate (boreal) coniferous forest ecosystems; ca 6 % support temperate deciduous forest ecosystems; ca 15 % support temperate grassland biomes (those regions that are too dry or otherwise unsuitable to permit the development of trees); and the remaining ca 27 % support tropical or subtropical biomes, whether forest or grassland, and comprising those regions that are never subjected to sub-freezing temperature.

The presumed integrated impact of a large-scale nuclear war during the growing season on the various systems, including both their flora and fauna, follows (Pittock *et al.*, 1985–1986, II, pp 85–111; Westing, 1980; cf. also Svirezhev *et al.*, 1985, pp 90–103 & 148–152):

a) in agricultural ecosystems (12 % of the life-supporting land areas), crop production would be prevented for at least one growing-season;
b) in tundra ecosystems (6 %), damage would be modest to minimal, but long-lasting;
c) in desert ecosystems (19 %), damage would be moderate and long-lasting;
d) in coniferous-tree ecosystems (15 %), damage would be moderate, with substantial recovery occurring over a period of some years;
e) in deciduous-tree ecosystems (6 %), damage would be rather severe, but recovery time would be approximately that of the coniferous systems;
f) in grass-dominated ecosystems (15 %), damage would be modest and recovery would occur rather quickly—that is, within a few years; and
g) in tropical ecosystems (27 %, but supporting perhaps two-thirds of the world's species of plants and animals), as already suggested, the expectable damage would be nothing short of catastrophic.

Thus, a large-scale nuclear war is sure to have many deleterious long-term effects on the biotas, some of them being predictable, but others less so. I have alluded to some but not all of these long-term effects. For example, weeds, pests, parasites, and disease organisms are in most instances more resistant to environmental stress (*e.g.*, radioactivity, UV-radiation, low temperature, or pollution) than are their victims, so that following a nuclear war these scourges would all tend to flourish in a relative sense (Hain, 1987; Westing, 1977, pp 20–22; Woodwell, 1970). Severe coastal storms might become far more prevalent in the months following a nuclear war than formerly, owing to the meeting and mixing of cold interior air masses with the less-cold water-buffered oceanic air masses, providing further stress to the biotas. The plant and animal extinctions that would occur as a result of the extraordinarily harsh wartime and post-war environmental conditions, will be sure to have unfavorable, though difficult-to-predict, secondary ramifications.

Perhaps of great importance, unfavorable synergisms are bound to emerge among the various horrendous stresses to living things which have already been described singly above. As one modest example of such a synergism I can mention that, in experiments with dogs, it has been found that when a seemingly benign level of radioactivity is presented together with a level of body burns from which most of the dogs are expected to recover, the combined effect, in fact, becomes lethal for the great majority of them (Brooks *et al.*, 1952).

In short, it becomes abundantly clear that an environmental disruption of the size (areal extent), intensity, and diversity of a nuclear war would be detrimental to the world's biotas in numerous direct and indirect fashions. It must really be stressed, however, that the predictability of the likelihood and extent of the various detrimental effects which have been described above varies enormously from one effect to the next. In addition other, as yet unknown, second-order effects seem certain to occur (McNaughton *et al.*, 1986).

7.8 Conclusion

Before ending, it is important to stress again that my outline of the potential long-term ecological consequences of nuclear war is based on extrapolations from very limited data. The profoundly horrible effects of the Hiroshima and Nagasaki attacks notwithstanding (Committee…, 1981), I believe that we—and especially our governments—have been lulled into a false sense of security and safety by the *relatively* benign nature of those terrible events. As macabre as this statement might sound, those two bombs were *comparatively* trivial in their impact: (a) owing to their small sizes; (b) owing to their isolated use; and (c) because as air bursts they generated almost no nuclear fallout.

The large-scale nuclear war to which this good earth could be subjected at any future moment, could be literally millions of times worse than the experiences of Hiroshima and Nagasaki in its immediate effects, and would additionally have countless serious long-term direct and indirect effects. Moreover, since the many adverse effects of enhanced UV-radiation and reduced temperatures could extend far beyond the confines of the combatant nations, the issue of nuclear war is now unequivocally a global one—of profound relevance not just to the two super-powers, but to all nations, and whether or not they possess nuclear weapons.

All in all, there is to me simply *no* conceivable end for which one could find a justification to employ these dreadful weapons—and thus *no* justification for even possessing them. I therefore hope fervently that ever-larger numbers of people in eastern and western Europe, in the USA, and elsewhere throughout the world, will come to recognize that nuclear war, and even the nuclear weapons themselves, must be renounced as the ultimate in human madness. Finally, let us only hope that, through us, our governments will learn and act in time.

References

ACDA. 1978. *Assessment of Frequently Neglected Effects in Nuclear Attacks*. Washington: *US Arms Control & Disarmament Agency*, Civil Defense Study Report No. 5, 21+7 pp.

Alcalay, G.H. 1980. Aftermath of Bikini. *Ecologist* (London) 10:346–351.

Aleksandrov, V.V., & Stenchikov, G.L. 1983. *On the Modelling of the Climatic Consequences of…Nuclear War*. Moscow: USSR Academy of Sciences Computing Centre, 21 pp.

Altman, P.L., & Dittmer, D.S. 1972–1974. *Biology Data Book*. 2nd edn. Bethesda, MD, USA: Federation of American Scientists for Experimental Biology, 2123 pp.

Barton, I.J., & Paltridge, G.W. 1984. 'Twilight at noon' overstated. *Ambio* (Stockholm) 13:59–51.

Bates, C.G., & Roeser, J., Jr. 1928. Light intensities required for growth of coniferous seedlings. *American Journal of Botany* (St Louis, MO, USA) 15:185–194.

Behar, A., *et al.* 1987. What if a nuclear warhead explodes over a one gigawatt nuclear power reactor in France? *Medicine & War* (London) 3:5–10.

Benarie, M. 1986. Volcanoes, weather, wine and winter (nuclear). *Science of the Total Environment* (Amsterdam) 50:191–196.

Bensen, D.W., & Sparrow, A.H. (eds). 1971. *Survival of Food Crops and Livestock in the Event of Nuclear War*. Washington: US Atomic Energy Commission Symposium Series No. 24, 745 pp.

Bergström, S., *et al.* 1987. *Effects of Nuclear War on Health and Health Services*. rev. edn. Geneva: *World Health Organization* Publication No. A40/11, 27+130 pp.

Bergthorsson, P. 1985. Sensitivity of Icelandic agriculture to climatic variations. *Climatic Change* (Dordrecht, Netherlands) 7:111–127.

Brinkman, A.W., & McGregor, J. 1983. Solar radiation in dense Saharan aerosol in northern Nigeria. *Quarterly Journal of the Royal Meteorological Society* (Bracknell, UK) 109:831–847.

Broecker, W.S. 1970. Man's oxygen reserves. *Science* (Washington) 168:1537–1538.

Brooks, J.W., *et al.* 1952. Influence of external body radiation on mortality from thermal burns. *Annals of Surgery* (Philadelphia) 136:533–545.

Caldwell, M.M. 1981. Plant response to solar ultraviolet radiation. In: *Encyclopedia of Plant Physiology*. new ser. New York: Springer-Verlag, vol. 12A, pp 170–197 (chap. 6).

Calkins, J., & Thordardottir, T. 1980. Ecological significance of solar UV radiation on aquatic organisms. *Nature* (London) 283:563–566.

Carrier, G.F., *et al.* 1985. *Effects on the Atmosphere of a Major Nuclear Exchange*. Washington: National Academy Press, 193 pp.

Catchpole, A.J.W. & Faurer, M.-A. (1983). Summer sea-ice severity in Hudson Strait, 1751–1870. *Climatic Change* (Dordrecht, Netherlands) 5:115–139.

Cess, R.D., *et al.* 1985. Climatic effects of large injections of atmospheric smoke and dust: a study of climate feedback mechanisms with one- and three-dimensional climate models. *Journal of Geophysical Research* (Washington) 90(D7):12937–12950.

Chazov, Y.I., *et al.* 1984. *Nuclear War: the Medical and Biological Consequences: Soviet Physicians' Viewpoint*. Moscow: Novosti Press Agency Publishing House, 239 pp.

Clark, W.C., *et al.* 1982. Carbon dioxide question: perspectives for 1982. In: Clark, W.C. (ed.). *Carbon Dioxide Review 1982*. Oxford: Clarendon Press, 469 pp.

Cogley, J.G. 1985. Hypsometry of the continents. *Zeitschrift für Geomorphologie*. new ser. (Berlin) 53 (supple.):1–48.

Committee for the Compilation of Materials on Damage Caused by the Atomic Bombs in Hiroshima and Nagasaki. 1981. *Hiroshima and Nagasaki: the Physical, Medical, and Social Effects of the Atomic Bombings* (transl. from the Japanese by Ishikawa, E., & Swain, D.L.). New York: Basic Books, 706 pp.

Cotton, W.R. 1985. Atmospheric convection and nuclear winter. *American Scientist* (Raleigh, NC, USA) 73:275–280.

Covey, C., *et al.* 1984. Global atmospheric effects of massive smoke injections from a nuclear war: results from general circulation model simulations. *Nature* (London) 308:21–25.

Covey, C., *et al.* 1985. 'Nuclear winter': a diagnosis of atmospheric general circulation model simulations. *Journal of Geophysical Research* (Washington) 90(D3):5615–5628.

Crutzen, P.J., *et al.* 1984. Atmospheric effects from post-nuclear fires. *Climatic Change* (Dordrecht, Netherlands) 6:323–364.

Daugherty, W., *et al.* 1985–1986). Consequences of 'limited' nuclear attacks on the United States. *International Security* (Cambridge, MA, USA) 10(4):3–45.

Din, A.M., & Diezi, J. 1984. *Nuclear War Effects in Switzerland*. Basel, Switzerland: Physicians for Social Responsibility, 43 pp + 33 figs.

Ehrlich, P.R., *et al.* 1983. Long-term biological consequences of nuclear war. *Science* (Washington) 222:1293–1300.

Ehrlich, R. 1986. We should not overstate the effects of nuclear war. *International Journal on World Peace* (New York) 3(3):31–43. cf. *ibid.* 3(3):43–46.

Faber, M., *et al.* 1979. *Ultraviolet Radiation*. Geneva: *World Health Organization* Environmental Health Criteria No. 14, 110 pp.

Fetter, S.A., & Tsipis, K. 1981. Catastrophic releases of radioactivity. *Scientific American* (New York) 244(4):33–39, 146.

Flavin, C. 1987. *Reassessing Nuclear Power: the Fallout from Chernobyl*. Washington: Worldwatch Institute Paper No. 75, 91 pp.

Fryxell, P.A. 1957. Mode of reproduction of higher plants. *Botanical Review* (Chicago) 23:135–233.

GAO. 1986. *Nuclear Winter: Uncertainties Surrounding the Long-term Effects of Nuclear War*. Washington: US General Accounting Office Report No. GAO/NSIAD-86–62, 55 pp.

Gerstl, S.A.W., *et al.* 1981. Biologically damaging radiation amplified by ozone depletions. *Nature* (London) 294:352–354.

Glasstone, S., & Dolan, P.J. 1977. *Effects of Nuclear Weapons*. 3rd edn. Washington: US Departments of Defense & Energy, 653 pp + slide-rule.

Golding, B.W., *et al.* 1986. Importance of local mesoscale factors in any assessment of nuclear winter. *Nature* (London) 319:301–303.

Greene, O., *et al.* 1985. *Nuclear Winter: the Evidence and the Risks*. Cambridge, UK: Polity Press, 216 pp.

Gromyko, A.A., *et al.* 1984. *Global Consequences of Nuclear War and the Developing Countries*. Moscow: Committee of Soviet Scientists for Peace, Against the Nuclear Threat, 42 pp.

Gunter, G., & Hildebrand, H.H. 1951. Destruction of fishes and other organisms on the south Texas coast by the cold wave of January 28–February 3, 1951. *Ecology* (Washington) 32:731–736.

Hain, F.P. 1987. Interactions of insects, trees and air pollutants. *Tree Physiology* (Victoria, Canada) 3:93–102.

Hare, F.K., *et al.* 1985. *Nuclear Winter and Associated Effects: a Canadian Appraisal of the Environmental Impact of Nuclear War.* Ottawa: Royal Society of Canada, 382 pp.

Harwell, M.A. 1984. *Nuclear Winter: the Human and Environmental Consequences of Nuclear War.* New York: Springer-Verlag, 179 pp.

Hecht, A.D., *et al.* 1985. *Interagency Research Report for Assessing Climatic Effects of Nuclear War.* Washington: US Office of Science Technology & Policy, 49+2+5 pp.

Hensel, H., *et al.* 1973. Homeothermic organisms. In: Precht, H., *et al.* (eds). *Temperature and Life.* Berlin: Springer-Verlag, 779 pp.

Hippel, F. von, & Cochran, T.B. 1986. Chernobyl: the emerging story: estimating long-term health effects. *Bulletin of the Atomic Scientists* (Chicago) 43(7):18–24.

Hobbs, P.V. 1985. 'Nuclear winter' calculations. *Science* (Washington) 228:648.

Hohenemser, C., *et al.* 1986. Chernobyl: an early report. *Environment* (Washington) 28(5):6–13, 30–43.

Holdsworth, G. 1986. Evidence for a link between atmospheric thermonuclear detonations and nitric acid. *Nature* (London) 324:551–553.

Holland, H.D. 1978. *Chemistry of the Atmosphere and Oceans.* New York: John Wiley, 351 pp.

Hoyt, J.B. 1958. Cold summer of 1816. *Annals of the Association of American Geographers* (Washington) 48:118–131.

Hutchinson, T.C. 1967. Comparative studies of the ability of species to withstand prolonged periods of darkness. *Journal of Ecology* (London) 55:291–299.

Johnson, G. 1980. Paradise lost. *Bulletin of the Atomic Scientists* (Chicago) 36(10):24–29.

Jokiel, P.L. 1980. Solar ultraviolet radiation and coral reef epifauna. *Science* (Washington) 207:1069–1071.

Jones, G.N. 1951. On the number of species of plants. *Scientific Monthly* (Washington) 72:289-294, 312.

Katz, A.M. 1982. *Life After Nuclear War: the Economic and Social Impacts of Nuclear Attacks on the United States.* Cambridge, MA, USA: Ballinger, 423 pp.

Keeling, C.D., *et al.* 1982. Measurements of the concentration of carbon dioxide at Mauna Loa Observatory, Hawaii. In: Clark, W.C. (ed.). *Carbon Dioxide Review 1982.* Oxford: Clarendon Press, 469 pp: pp 377–385.

Korstian, C.F. 1921. Effect of a late spring frost upon forest vegetation in the Wasatch Mountains of Utah. *Ecology* (Washington) 2(1):47–52.

Kruger, C.H., Jr, *et al.* 1982. *Causes and Effects of Stratospheric Ozone Reduction: an Update.* Washington: National Academy Press, 339 pp.

Larsson, T. 1981. Environmental effects of nuclear war. In: Barnaby, W. (ed.). *War and Environment.* Stockholm: Royal Ministry of Agriculture, Environmental Advisory Council, 154 pp: pp 34–150.

Laudien, H. 1973. Poikilothermic organisms: animals: activity, behavior, etc. In: Precht, H. *et al.* (eds). *Temperature and Life.* Berlin: Springer-Verlag, 779 pp: pp 441–469.

Lee, R.E., Jr, *et al.* 1987. Rapid cold-hardening process in insects. *Science* (Washington) 238:1415–1417.

Lyons, J.M. 1973. Chilling injury in plants. *Annual Review of Plant Physiology* (Palo Alto, CA, USA) 24:445–466.

MacDonald, L. 1984. Wound in the world. *Asiaweek* (Hong Kong) 10(28):32–38, 43–49.

McNaughton, S.J., *et al.* 1986. Ecological consequences of nuclear war. *Nature* (London) 321:483–487.

Malingreau, J.P., *et al.* 1985. Remote sensing of forest fires: Kalimantan and North Borneo in 1982–83. *Ambio* (Stockholm) 14:314–321.

Malone, R.C., *et al.* 1986. Nuclear winter: three-dimensional simulations including interactive transport, scavenging, and solar heating of smoke. *Journal of Geophysical Research* (Washington) 91(Dl):1039–1053.

Manley, G. 1974. Central England temperatures: monthly means 1659 to 1973. *Quarterly Journal of the Royal Meteorological Society* (Bracknell, UK) 100:389–405.

Milne, D.H., & McKay, C.P. 1982. Response of marine plankton communities to a global atmospheric darkening. In: Silver, L.T., & Schultz, P.H. (eds). *Geological Implications of Impacts of Large Asteroids and Comets on the Earth.* Boulder, CO, USA: Geological Society of America, Special Paper No. 190, 528 pp: pp 297–303.

Mitchell, B.R. 1981. *European Historical Statistics 1750–1975.* 2nd edn. London: Macmillan, 868 pp.

Neftel, A., *et al.* 1985. Evidence from polar ice-cores for the increase in atmospheric CO_2 in the past two centuries. *Nature* (London) 315:45–47.

Nier, A.O.C., *et al.* 1975. *Long-term Worldwide Effects of Multiple Nuclear-weapons Detonations.* Washington: US National Academy of Sciences, 213 pp.

Paltridge, G.W., & Hunt, G.E. 1984. Three-dimensional climate models in perspective. *Ambio* (Stockholm) 13:387–388.

Park, C. 1978. Nitric oxide production by Tunguska meteor. *Acta Astronautica* (Elmsford, NY, USA) 5:523–542.

Parker, J. 1963. Cold resistance in woody plants. *Botanical Review* (Chicago) 29:123–201.

Pearman, G.I., *et al.* 1986. Evidence of changing concentrations of atmospheric CO_2, N_2O and CH_4 from air bubbles in Antarctic ice. *Nature* (London) 320:248–250.

Penner, J.E. 1986. Uncertainties in the smoke source term for 'nuclear winter' studies. *Nature* (London) 324:222–226.

Petersen, R.C., Jr, *et al.* 1986. Assessment of the impact of the Chernobyl reactor accident on the biota of Swedish streams and lakes. *Ambio* (Stockholm) 15:327–331.

Peterson, J.T., & Drury, L.D. 1967. Reduced values of solar radiation with occurrence of dense smoke over the Canadian tundra. *Geographical Bulletin* (Ottawa) 9:269–271.

Pittock, A.B., *et al.* 1985–1986. *Environmental Consequences of Nuclear War. I. Physical and Atmospheric Effects. II. Ecological and Agricultural Effects.* Chichester, UK: John Wiley & Sons, 359+523 pp.

Post, J.D. 1974. Study in meteorological and trade cycle history: the economic crisis following the Napoleonic wars. *Journal of Economic History* (New York) 34:315–349.

Precht, H. 1973. Poikilothermic organisms: animals: limiting temperatures of life functions. In: Precht, H., *et al.* (eds). *Temperature and Life.* Berlin: Springer-Verlag, 779 pp: pp 400–440.

Ramberg, B. 1984. *Nuclear Power Plants as Weapons for the Enemy: an Unrecognized Peril.* 2nd edn. Berkeley, CA, USA: University of California Press, 193 pp.

Rasmussen, K.L., *et al.* 1984. Nitrate in the Greenland ice-sheet in the years following the 1908 Tunguska event. *Icarus* (New York) 58:101–108.

Rathjens, G.W., & Siegel, R.H. 1984–1985. Nuclear winter: strategic significance. *Issues in Science & Technology* (Washington) 1(2):123–128.

Rathjens, G.W., & Siegel, R.H. 1985. Review of '*The Cold and the Dark*' by P.R. Ehrlich *et al.* *Survival* (London) 27:43–44.

Raven, P.H. 1976. Ethics and attitudes. In: Simmons, J.B., *et al.* (eds). *Conservation of Threatened Plants.* New York: Plenum Press, 336 pp: pp 155–179.

Raven, P.H., *et al.* 1981. *Biology of Plants.* 3rd edn. New York: Worth, 686 pp.

Regal, P.J. 1982. Pollination by wind and animals: ecology of geographic patterns. *Annual Review of Ecology & Systematics* (Palo Alto, CA, USA) 13:497–524.

Robock, A. 1984. Snow and ice feedbacks prolong effects of nuclear winter. *Nature* (London) 310:667–670.

Rotblat, J. 1981. *Nuclear Radiation in Warfare.* London: Taylor & Francis, 149 pp.

Sakai, A. & Otsuka, K. 1970. Freezing resistance of alpine plants. *Ecology* (Washington) 51:665–671.

Sakai, A., & Weiser, C.J. 1973. Freezing resistance of trees in North America with reference to tree regions. *Ecology* (Washington) 54:118–126.

Schneider, S.H. 1987. Climate modeling. *Scientific American* (New York) 256(5):72–81, 120.

Schultz, V., & Whicker, F.W. 1985. Ionizing radiation and nuclear war: review of deliberations on ecological impacts. *Critical Reviews in Environmental Control* (Boca Raton, FL, USA) 15:417–427.

Sear, C.B., *et al.* 1987. Global surface-temperature responses to major volcanic eruptions. *Nature* (London) 330:365–367.

Seitz, R. 1986. Siberian fire as 'nuclear winter' guide. *Nature* (London) 323:116–117.

Sharfman, P., *et al.* 1979. *Effects of Nuclear War.* Washington: US Congress Office of Technology Assessment Publ. No. OTA-NS-89, 151 pp.

Shetler, S.G., & Skog, L.E. 1978. *Provisional Checklist of Species for* Flora North America. rev. edn. St Louis, MO, USA: Missouri Botanical Garden, 199 pp.

Shinn, E.A. 1976. Coral reef recovery in Florida and the Persian Gulf. *Environmental Geology* (New York) 1:241–254.

Shirley, H.L. 1929. Influence of light intensity and light quality upon the growth of plants. *American Journal of Botany* (St Louis, MO, USA) 16:354–390 + plates XXVIII-XXXII.

Shostakovitch, V.B. 1925. Forest conflagrations in Siberia: with special reference to the fires of 1915. *Journal of Forestry* (Washington) 23:365–371.

Singer, S.F. 1984. Is the 'nuclear winter' real? *Nature* (London) 310:625.

Singer, S.F. 1985. On a 'nuclear winter'. *Science* (Washington) 227:356.

Small, R.D., & Bush, B.W. 1985. Smoke production from multiple nuclear explosions in nonurban areas. *Science* (Washington) 229:465–469.

Smith, J.V. 1985. Review of '*The Cold and the Dark*' by P.R. Ehrlich *et al. Bulletin of the Atomic Scientists* (Chicago) 41(1):49–51.

Solomon, F., & Marston, R.Q. (eds). 1986. *Medical Implications of Nuclear War.* Washington: National Academy Press, 619 pp.

Stenchikov, G. 1985. Climatic consequences of nuclear war. In: Velikhov. Y. (ed.). *The Night After...: Scientists' Warning.* Moscow: Mir, 165 pp: pp 53–82.

Stommel, H., & Stommel, E. 1979. Year without a summer. *Scientific American* (New York) 240(6):176–86, 202.

Stothers, R.B. 1984. Great Tambora eruption in 1815 and its aftermath. *Science* (Washington) 224:1191–1198.

Svirezhev, Y.M., *et al.* 1985. *Ecological and Demographic Consequences of Nuclear War.* Moscow: USSR Academy of Sciences Computer Center, 267 pp.

Teller, E. 1984. Widespread after-effects of nuclear war. *Nature* (London) 310:21–24.

Teramura, A.H. 1983. Effects of ultraviolet-B radiation on the growth and yield of crop plants. *Physiologia Plantarum* (Copenhagen) 58:415–427.

Thompson, S.L. 1985. Global interactive transport simulations of nuclear war smoke. *Nature* (London) 317:35–39.

Thompson, S.L., & Schneider, S.H. 1985–1986. Nuclear winter reappraised. *Foreign Affairs* (New York) 64:981–1005; 65:163–178.

Thunborg, A.I. *et al.* 1981. *Comprehensive Study on Nuclear Weapons.* New York: United Nations Centre for Disarmament Study Series No. 1, 172 pp.

Toumey, J.W,. & Korstian, C.F. 1947. *Foundations of Silviculture upon an Ecological Basis.* 2nd edn, rev. New York: John Wiley 468 pp.

Tukey, J.W., *et al.* 1979. *Protection Against Depletion of Stratospheric Ozone by Chlorofluorocarbons.* Washington: US National Academy of Sciences, 392 pp.

Turco, R.P., *et al.* 1981. Tunguska meteor fall of 1908: effects on stratospheric ozone. *Science* (Washington) 214:19–23.

Turco, R.P., *et al.* 1982. Analysis of the physical, chemical, optical, and historical impacts of the 1908 Tunguska meteor fall. *Icarus* (New York) 50:1–52.

Turco, R.P., *et al.* 1983. Nuclear winter: global consequences of multiple nuclear explosions. *Science* (Washington) 222:1283–1292.

Velikhov, Y. (ed.). 1985. *The Night After...: Scientists' Warning*. Moscow: Mir, 165 pp.

Warren, S.G., & Wiscombe, W.J. 1985. Dirty snow after nuclear war. *Nature* (London) 313:467–470.

Watts, J.A. 1982. Carbon dioxide question: data sampler. In: Clark, W.C. (ed.). *Carbon Dioxide Review 1982*. Oxford: Clarendon Press, 469 pp: pp 431–469.

Weinberger, C.W. 1985–1986. *Potential Effects of Nuclear War on the Climate*. Washington: US Department of Defense, 17+5 pp.

Westing, A.H. 1977. *Weapons of Mass Destruction and the Environment*. London: Taylor & Francis, 95 pp.

Westing, A.H. 1978. Neutron bombs and the environment. *Ambio* (Stockholm) 7(3):93–97.

Westing, A.H. 1980. *Warfare in a Fragile World: Military Impact on the Human Environment*. London: Taylor & Francis, 249 pp.

Westing, A.H. 1981. Environmental impact of nuclear warfare. *Environmental Conservation* (Cambridge, UK) 8(4):269–273.

Westing, A.H. 1982. Environmental consequences of nuclear warfare. *Environmental Conservation* (Cambridge, UK) 9(4):269–272.

Westing, A.H. 1985a. Nuclear winter: a bibliography. *SIPRI Yearbook* (Oxford) 1985:126–129.

Westing, A.H. 1985b. Review of '*Nuclear Winter*' by M.A. Harwell. *Environment* (Washington) 27(4):28–29.

Westing, A.H. 1986. Review of '*Environmental Consequences of Nuclear War*' by A.B. Pittock *et al. Environmental Conservation* (Cambridge, UK) 13(3):281–282.

Wexler, H. 1950. Great smoke pall: September 24–30, 1950. *Weatherwise* (Philadelphia) 3:129–134, 142.

Woodwell, G.M. 1970. Effects of pollution on the structure and physiology of ecosystems. *Science* (Washington) 168:429–433.

Worrest, R.C. 1983. Impact of solar ultraviolet-B radiation (290–320 nm) upon marine microalgae. *Physiologia Plantarum* (Copenhagen) 58:28–34.

Worrest, R.C., *et al.* 1981a. Impact of UV-B radiation upon estuarine microcosms. *Photochemistry & Photobiology* (London) 33:861–867.

Worrest, R.C., *et al.* 1981b. Sensitivity of marine phytoplankton to UV-B radiation: impact upon a model ecosystem. *Photochemistry & Photobiology* (London) 33:223–227.

Chapter 8
Protecting the Environment in War: Legal Constraints

Note : *Our cultural norms, both social and environmental with respect to both peace and war, have in modern times been undergoing a slow progression for the better (#380).*[1] *Thus the truth of the old aphorism that 'All is fair in love and war' is being slowly eroded, at least in principle if not so robustly in practice. Resolutions by the United Nations General Assembly, decrees by the United Nations Security Council, expansions of the Law of War (International Humanitarian Law as well as the related International Arms Control and Disarmament Law), and other factors all point in that direction—including a number relative to the specific subject of this chapter, viz., as these cultural norms refer to protection of the environment from military actions. Surprisingly (and disappointingly), however, there is essentially no evidence of this welcome trend of increased wartime environmental protection to be found within either International Environmental Law or International Human Rights Law. Even sadder, there is even an anti-environmental component discernible within International Maritime Law (#233).*

What is reproduced below (#311) presents the existing wartime protections of the environment deriving from international law. The protections, as they now stand, are applicable largely to international armed conflicts, and thus substantially less so to non-international (internal) armed conflicts. But all of the existing constraints, of course, serve on the one hand to be proscriptive, and on the other to be normative. Specific reference is made below to the Second Indochina War of

[1] The numbered references are provided in Chap. 3.

A. H. Westing, *Arthur H. Westing*, SpringerBriefs on Pioneers in Science and Practice 1, DOI: 10.1007/978-3-642-31322-6_8,

1961–1975 (cf. *Chap. 4) as well as to the Gulf War of 1991 (* cf. *Chap. 5). The subsequent presentation* (cf. *Chap. 9), in turn, describes how the existing legal proscriptions presented in this chapter actually play out nationally.*[2]

Abstract The primary question examined in this study is the extent to which international law can be expected to mitigate environmental disruption in times of warfare, whether interstate or intrastate (internal). Approaches to protecting the environment from military damage that have legal precedents include: (a) remaining at peace; (b) establishing zones of peace; (c) limiting certain weapons; (d) limiting certain means of warfare; and (e) limiting damage to natural resources. Of the various bodies of international law, neither International Environmental Law nor International Human Rights Law seems directly applicable to the question at hand, whereas International Humanitarian Law (the Law of War), including International Arms Control and Disarmament Law, is. Three relatively recent multilateral treaties can be singled out for their specific relevance to environmental disruption during wartime: (a) 1977 Protocol I on International Armed Conflicts (UNTS 17512); (b) 1980 Land Mine Protocol II of the 1980 Inhumane Weapon Convention [regarding international armed conflicts] (UNTS 22495); and (c) 1977 Protocol II on Non-international Armed Conflicts (UNTS 17513). The first of these has explicitly expanded International Humanitarian Law to encompass environmental concerns per se. A consideration of these three instruments, together with other components of International Humanitarian Law which provide incidental protection to the environment, suggests that existing constraints are about as stringent as is currently feasible, given the state of the underlying cultural norms throughout the world. It is concluded that a state (nation) which becomes party to an International Humanitarian Law treaty does so in good faith and can generally be expected to comply with its strictures. It is noted that a preponderance of the numerous states non-parties to important relevant instruments suffer from some combination of poverty, lack of human or other development, and paucity of basic freedoms. Increased treaty participation, and firmer expectations of compliance, will depend upon a combination of widespread military and civil education to nurture relevant underlying cultural norms on the one hand, and the alleviation of poverty and spread of democratization and integrity on the other. Finally, it is recommended (a) that a treaty be adopted that would prohibit the use in war of nuclear weapons, and (b) that natural heritage sites of outstanding universal value be designated as demilitarized zones.

---

[2] Reproduced from: Gleditsch, N.P., *et al.* (eds). *Conflict and the Environment.* Dordrecht, Netherlands: Kluwer Academic Publishers, 598 pp: pp 535–553 (Chap. 32); 1997 with the original title *"Environmental Protection from Wartime Damage: The Role of International Law"* by permission of the Springer Verlag, the copyright holder, on 14 March 2012. The author is pleased to acknowledge useful suggestions from Richard A. Falk (Princeton), Jozef Goldblat (Geneva), Jean Perrenoud (Geneva), Nico J. Schrijver (The Hague), Christopher D. Stone (Los Angeles), Wil D. Verwey (Groningen), and Carol E. Westing (Putney).

8.1 The Issue

There is no question that warfare can be more or less seriously disruptive of the environment. A substantial fraction of such environmental disruption is unintentional (ancillary, incidental, collateral), but some of it is intentional. Although no separation need be made here between declared warfare and undeclared warfare, it is necessary in the present context to distinguish between interstate warfare and intrastate (internal, civil) warfare. Moreover, as to the environment, the present examination is by and large limited to the terrestrial and marine environments (thus for the most part omitting considerations of the atmosphere and of celestial bodies and outer space).

The primary question examined in this study is the extent to which international law can be expected to mitigate environmental disruption in times of warfare, bearing in mind both unintentional and intentional environmental disruption and both interstate and intrastate warfare. Examined first is the applicable international law. This is followed by an examination of some of the characteristics of states that tend to differ between states parties and states non-parties to the relevant treaties. Special attention is paid to three relatively recent highly pertinent additions to international law, two of these treaties dealing specifically with interstate warfare and the other specifically with intrastate warfare. There is a particular need to distinguish between interstate and intrastate warfare because the bulk of applicable international law is relevant only to interstate warfare, whereas most wars in recent decades have not been interstate but intrastate.

8.2 Applicable International Law

8.2.1 Approaches to Protecting the Environment from Military Damage

Of the various legal approaches of a general nature to protecting the environment from military damage, five are singled out here:

(a) *Remaining at peace:* The first approach to protecting the environment from warfare would, of course, be to put an end to war. However, this is a utopian notion since—endless rhetoric aside—warfare is one of the defining characteristics of our species. Indeed, there has almost certainly never been even a single day of peace throughout the long sweep of human history. Merely since 1929, the year in which the then almost universally adopted 1928 Treaty for the Renunciation of War (*LNTS* 2137) entered into force (Westing, 1990), several hundred wars have been fought (Sivard, 1996, pp 18–19; Westing, 1982). The more recently adopted 1945 *United Nations Charter*, another almost universally adopted multilateral treaty (185 of the 192 current states, or 96 %, being parties), although not actually prohibiting interstate warfare, contains significant constraints on warfare, also without demonstrable effect. Moreover, neither of these instruments pertains to intrastate warfare.

(b) *Establishing zones of peace:* A second approach to protecting the environment from warfare, or at least some greater or lesser portions of the environment, is to establish areas from which warfare is to be excluded. The modestly popular 1959 Antarctic Treaty (*UNTS* 5778) is, *inter alia*, meant to keep the land area of Antarctica at peace (Article 1). The Svalbard archipelago in the Arctic Ocean (*LNTS* 41, Article 9), the Åland archipelago in the Baltic Sea (*LNTS* 255, Articles 3 etc.), and several lesser areas are provided for in similar fashion. Moreover, almost 100 natural heritage sites of outstanding universal value are recognized via the quite widely adopted 1972 World Heritage Convention (*UNTS* 15511) in which the states parties agree not to take any deliberate measures, including armed conflict, that might harm such sites located on the territory of other states parties (Articles 11.4 & 13.1). It might also be mentioned that several countries have established a status of 'neutrality', a major aim of which is to avoid the depredations of warfare (Westing, 1990). Finally, a number of regions have been designated by the regional states as being—if not a warfare-free zone—at least a nuclear-weapon-free zone (including the Latin American plus Caribbean region [*UNTS* 9068] and the South Pacific region [*UNTS* 24592]).

(c) *Limiting certain weapons:* A third approach to protecting the environment from warfare, or at least from some potential aspects of warfare, is to prohibit or limit the use of environmentally disruptive weapons. Although the several extant examples of this approach were motivated by social rather than environmental concerns, they nonetheless are environmentally beneficial. One well known illustration is the widely adopted 1925 Geneva Protocol on Chemical and Bacteriological Warfare (*LNTS* 2138), which prohibits the states parties from using such agents in interstate warfare among themselves (although, regrettably, about half of the states parties have taken it upon themselves to reserve the option to retaliate in kind) (Westing, 1985b). Another is the not so widely adopted 1980 Inhumane Weapon Convention (*UNTS* 22495), one optional protocol of which provides for certain restraints on the use of land mines (Protocol II, hereinafter referred to as the '1980 Land Mine Protocol'). Another of its protocols includes a modest constraint on attacking forests or other plant cover with incendiary weapons (Protocol III, Article 2). By way of an aside, it is a tragedy—though an instructive one—that no example can as yet be offered of a treaty in force that would prohibit the use in warfare of nuclear weapons, clearly the most socially and environmentally damaging instruments of war now available in the arsenals of the major powers (Westing, 1987; 1989b).

(d) *Limiting certain means of warfare:* A fourth approach to protecting the environment from warfare is to prohibit or limit environmentally disruptive means of warfare. For example, the quite widely adopted 1977 Protocol I on International Armed Conflicts (*UNTS* 17512), *inter alia*, provides for constraints, again for social rather than environmental reasons, on attacking agricultural areas (Article 54.2) or dams or nuclear electrical generating stations that would release so-called dangerous forces (Article 56.1). However, specifically with reference to the natural environment, the states parties to Protocol I, agree, among themselves, not to use means of warfare that may be expected to cause the natural environment widespread, long-lasting, and severe damage (Articles 35.3 & 55.1)—a novel addition, it must be stressed, to International Humanitarian Law. Also to be mentioned in this category is the rather weak and not widely adopted 1977 Environmental Modification Convention (*UNTS* 17119), which is meant to restrict environmental manipulations for hostile interstate purposes (Westing, 1984; 1993a).

(e) *Limiting damage to natural resources:* A fifth approach to protecting the environment from warfare is to prohibit or limit destruction, seizure, or over-exploitation of natural resources. Destruction or seizure of enemy property is constrained by 1899 Hague Convention II and 1907 Hague Convention IV on the Laws and Customs of War on Land (both Conventions, Annex Article 23.g) as well as by 1949 Geneva Convention IV for the Protection of Civilians in Time of War (*UNTS* 973, Article 53). Over-exploitation, that is, exploitation in an environmentally damaging fashion, is limited by 1899 Hague Convention II and 1907 Hague Convention IV, which, *inter alia*, prohibit the non-usufructory exploitation by an occupying power of forests and agricultural works in enemy territory (both Conventions, Annex Article 55).

8.2.2 Potentially Applicable Bodies of International Law

International law is usually thought of in terms of a number of somewhat arbitrary and overlapping bodies of law. Three of these bear at least brief discussion, dwelling here only upon the multilateral treaties that comprise them (and thus excluding for present purposes any arbitral decisions and other possible components, including such subsidiary sources of international law as non-binding resolutions and unilateral declarations):

(a) *International Environmental Law:* To begin with, International Environmental Law is a rapidly expanding body of law, with dozens of major multilateral treaties now in force, virtually all of them having originated since 1950 (Westing, 1994b, Table 8). Although these treaties are in principle applicable during both peacetime and wartime—being for the most part silent on that distinction—regrettably, it seems to be widely accepted implicitly among the states parties that this body of law is operative only in times and places of peace. (The 1969 Vienna Convention on the Law of Treaties [*UNTS* 18232] does not provide an answer to this sensitive issue [cf. Article 73]). Indeed, of the major environmental protection treaties, whether non-domain specific or terrestrially oriented, only one—the 1972 World Heritage Convention—specifies its applicability to warfare (cf. above). And as to the major marine environmental protection treaties, most go so far as to specifically exempt warships and other naval vessels from their strictures, even in times and places of peace (the principle of 'sovereign immunity') (*e.g.*, *UNTS* 4714, Article 19; *UNTS* 6465, Article 8; *UNTS* 15749, Article 7.4; *UNTS* 22484, Article 3.3; *UNTS* 31363, Articles 236, etc.) (Westing, 1992a).

(b) *International Human Rights Law:* A second body of law to consider is International Human Rights Law. The number of major multilateral treaties in this category—now about two dozen in number—has more than doubled since 1950 (Westing, 1994b, Table 8). These treaties are meant to be applicable during both peace and war, and not merely to the states parties among themselves; nor is it legally possible to denounce them. However, they are of only peripheral interest in the present context because they have very little to say about environmental protection per se. Perhaps the closest this body of law comes to environmental protection, whether during peacetime or wartime, is through the 1966 Covenant on Civil and Political Rights (*UNTS* 14668). It is via this Covenant that the states parties have accepted that 'every human being has the inherent right to life' (Article 6), which—it can be suggested—implies a right to an environment adequate for its realization (Westing, 1993c).

(c) *International Humanitarian Law:* The third, and most important, body of law to consider is International Humanitarian Law, the terminological successor to the 'Law(s) of War'. International Arms Control and Disarmament Law is here included within this category. This overall body of law derives almost exclusively from social concerns, but nevertheless in the process provides for some substantive environmental safeguards. Moreover, in what must be recognized as a major innovation, 1977 Protocol I has, as was already alluded to earlier, explicitly expanded the notion of humanitarian concerns to include environmental concerns in their own right. As another point to emphasize for this large body of law, it is crucial to stress that the vast bulk of it is applicable only to interstate warfare, although one relatively modest instrument—1977 Protocol II on Non-international Armed Conflicts (*UNTS* 17513)—does deal directly with intrastate warfare. Indeed, it has been suggested that the *United Nations General Assembly* urge individual states to incorporate at least the four 1949 Geneva Conventions into domestic law in order to make them applicable internally (Lopez, 1994).

Before continuing with specifically designated environmental considerations, it will be useful to point out that one of the basic premises of International Humanitarian Law is that the right of belligerents to choose methods of warfare is not unlimited. This fundamental principle of restraint appears in essence in 1899 Hague Convention II (Annex Article 22), 1907 Hague Convention IV (Annex

Article 22), 1977 Protocol I (Article 35.1), and the 1980 Inhumane Weapon Convention (Preamble). An equally important, though perhaps even more nebulous, principle of this body of law (one known after its author, Feodor Martens) is that those military actions not precisely regulated are to be controlled by the principles of humanity and the dictates of the public conscience. This latter restraint is included in 1907 Hague Convention IV (Preamble), the 1945 International Court of Justice Statute (Article 38), the four 1949 Geneva Conventions (respectively, Articles 63, 62, 142, & 158), 1977 Protocol I (Article 1.2), 1977 Protocol II (Preamble), and the 1980 Inhumane Weapon Convention (Preamble). It has been suggested that the Martens clause can be translated into four basic principles of wartime limitation: (**a**) the principle of necessity; (**b**) the principle of discrimination; (**c**) the principle of proportionality; and (**d**) the principle of humanity (Falk, 1975).

8.3 Treaty Non-Parties Versus Treaty Parties

Three relatively recent multilateral treaties are singled out here for special examination owing to their specific relevance to environmental disruption by warfare (two by interstate warfare and one by intrastate warfare).

8.3.1 Interstate Warfare

1977 Protocol I, *inter alia*, commits its parties, among themselves, to take care to protect the natural environment in interstate warfare against widespread, long-lasting, and severe damage (Articles 35.3 & 55.1); and further provides for certain constraints on the destruction of agricultural areas (Article 54.2), and also on the release of dangerous forces (Article 56.1). About three-fourths of all states have so far joined this remarkable treaty, a high level of formal acceptance. Moreover, states have become parties more or less irrespective of their national wealth, industrialization, human development, or political status (cf. Table 8.1). Oddly enough, states that have been involved in major warfare in recent decades have been somewhat less apt to become parties than states without such experience.

A most interesting feature of 1977 Protocol I is the opportunity provided by its Article 90 for states parties to submit on a compulsory basis to ('accept the competence of') an International Fact-finding Commission empowered to look into allegations of serious breaches of this treaty (as well as of such breaches of the four 1949 Geneva Conventions). However, this optional verification procedure has so far been accepted on a compulsory basis by only about one-third of the states parties, that is, by only about one-fourth of all states (cf. Table 8.1). It could be suggested that this low latter level of acceptance represents at least in part an indication of the degree of true commitment to the strictures imposed by the treaty,

Table 8.1 Treaty participation

State Category (Number)	**Interstate** (%)			**Intrastate** (%)
	77 Pr. I (143)	**Art. 90** (47)	**80 Mine Pr.** (55)	**77 Pr. II** (134)
All States (192)	74	24	29	70
Permanent UNSC (5)	40	20	100	60
NATO (16)	75	69	69	81
In Major Wars (47)	64	13	28	57
In Minor or No Wars (145)	78	28	29	74
Heavily Mined (14)	43	14	14	29
Lightly or Not Mined (178)	77	25	30	73
GNP/Caput:				
High (36)	69	53	61	67
Upper Middle (29)	86	31	31	79
Lower Middle (68)	74	19	21	69
Low (59)	73	10	17	68
Industrialized (27)	81	78	89	85
In Between (117)	74	18	23	68
Least Developed (48)	73	10	8	67
HDI: Best: 0.85-0.95 (40)	80	52	60	80
0.75–0.85 (23)	77	20	27	60
0.55–0.75 (29)	78	14	14	78
0.35–0.55 (24)	75	9	16	75
Worst: 0.15–0.35 (18)	75	8	8	33
Free (76)	76	39	46	76
Partly Free (62)	71	16	21	69
Not Free (53)	75	13	13	6
Low Corruption (20)	65	65	90	70
Medium Corruption (15)	80	47	47	80
High Corruption (18)	61	11	44	61

(continued)

Table 8.1 (coninued)

Notes

(a) *Treaty abbreviations:* '77 Pr. I' = 1977 Protocol I on International Armed Conflicts (*UNTS* 17512). 'Art. 90' = 1977 Protocol I, Article 90, International Fact-finding Commission. '80 Mine Pr.' = 1980 Inhumane Weapon Convention (*UNTS* 22495), Protocol II on the Use of [Land] Mines, Booby-traps and Other Devices. '77 Pr. II' = 1977 Protocol II on Non-international Armed Conflicts (*UNTS* 17513).

(b) *States Considered:* The 192 *de facto* states of the world are the 185 members of the United Nations plus Kiribati, Nauru, Switzerland, Tonga, Taiwan, Tuvalu, and Vatican City.

(c) *War Participation:* The 47 states in major wars are those states that have been involved in wars, whether interstate or intrastate (internal), since 1945 that have resulted in at least 30 thousand fatalities, whether military or civilian (Sivard, 1996, pp 18-19; Westing, 1982). They are: Afghanistan, Algeria, Angola, Argentina, Bangladesh, Bosnia-Herzegovina, Burundi, Cambodia, Cameroon, China, Colombia, Croatia, Egypt, El Salvador, Eritrea, Ethiopia, France, Greece, Guatemala, India, Indonesia, Iran, Iraq, Israel, North Korea, South Korea, Kuwait, Laos, Lebanon, Liberia, Mozambique, Nicaragua, Nigeria, Philippines, Portugal, Russia, Rwanda, Somalia, Sri Lanka, Sudan, Tajikistan, Uganda, United Kingdom, USA, Viet Nam, Yemen, and Zaïre.

(d) *Mined States:* The 14 heavily mined states—those with at least 1 million residual land mines—are Afghanistan, Angola, Bosnia-Herzegovina, Cambodia, Croatia, Eritrea, Ethiopia, Iran, Iraq, Kuwait, Mozambique, Serbia-Montenegro, Somalia, and Sudan (USDOS, 1993; 1994).

(e) *Wealth (GNP) of States:* The 192 states are segregated annually by the *International Bank for Reconstruction & Development [World Bank]* (Washington) into the following four wealth categories, using gross national product (GNP) per caput (1993 data): 'High', US$ 8626 or more; 'Upper middle', US$ 2786–8625; 'Lower middle', US$ 696-2785; and 'Low', US$ 695 or less (IBRD, 1995, pp 248-249; with 4 missing values from CIA, 1995).

(f) *State Development:* The 27 industrialized states are: Australia, Austria, Belgium, Canada, Czech Republic, Denmark, Finland, France, Germany, Hungary, Iceland, Ireland, Italy, Japan, Luxembourg, Netherlands, New Zealand, Norway, Poland, Portugal, Romania, Russia, South Africa, Sweden, Switzerland, United Kingdom, and USA. The 48 'least developed countries'—as designated by the *United Nations General Assembly* (New York) on the basis of per-caput income, literacy rate, and industrial development—are: Afghanistan, Angola, Bangladesh, Benin, Bhutan, Burkina Faso, Burundi, Cambodia, Cape Verde, Central African Republic, Chad, Comoros, Djibouti, Equatorial Guinea, Eritrea, Ethiopia, Gambia, Guinea, Guinea-Bissau, Haiti, Kiribati, Laos, Lesotho, Liberia, Madagascar, Malawi, Maldives, Mali, Mauritania, Mozambique, Myanmar, Nepal, Niger, Rwanda, Samoa (Western), São Tomé-Principe, Sierra Leone, Solomon Islands, Somalia, Sudan, Tanzania, Togo, Tuvalu, Uganda, Vanuatu, Yemen, Zaïre, and Zambia.

(g) *State Human Development:* The 'human development' index (HDI), which in theory ranges from '1' (best) to '0' (worst), is compiled annually by the United Nations Development Programme (New York) on the basis of life expectancy at birth, adult literacy rate, school enrollment, and gross domestic product per caput (UNDP, 1995, pp 155–157,134–135). The states are here divided, from best to worst, into five categories (cf. Table 8.1 above).

(h) *State Freedom:* A 'freedom' index for states is compiled annually by *Freedom House* (New York) in a range from '1' (best) to '7' (worst) on the basis of an examination of political rights and civil liberties (Kaplan, 1996, pp 536–537,541). The states are divided by Freedom House into the following three categories: 'Free', 1.0–2.5; 'Partly free', 3.0–5.0; and 'Not free', 5.5–7.0.

(i) *State Corruption:* A governmental corruption index for states is compiled annually by *Transparency International* (Berlin), in theory ranging from '10' (not corrupt) to '0' (fully corrupt), the value for each state being determined on the basis of at least four international business surveys (Lambsdorff, 1996, p. 3). Regrettably, values are provided for only 53 (28 %) of the 192 current states. The states are here divided into 'Low' (6.67–10), 'Medium' (3.34–6.66), and 'High' (0–3.33) levels of perceived governmental corruption (lack of integrity) (cf. Table 8.1 above).

a matter discussed further below. It is thus of interest to recognize the clear positive correlation between acceptance to date of Article 90 and the levels of wealth, industrialization, human development, political freedom, and level of governmental integrity (non-corruption) of a state.

A characteristic of many wars today (both interstate and intrastate) is the heavy reliance on land mines, for both defensive and offensive purposes. This is a tragedy because most land mines do not detonate at the time they are emplaced, but remain hidden in the rural environment after the hostilities cease, only to become a terrible long-term postwar hazard to the civilian population and a great impediment to postwar reconstruction (Westing, 1985a). The 1980 Land Mine Protocol (*UNTS* 22495) commits its parties, among themselves, to certain restrictions during interstate warfare on the use of these pernicious additions to the environment. Despite the rather modest constraints associated with this treaty, less than one-third of all states have so far seen fit to join it (cf. Table 8.1). Moreover, wartime experience in recent decades has had no apparent influence on acceptance level; and the dozen or so heavily mined states (for present purposes, those currently infested with a million or more residual land mines) inexplicably seem to be avoiding the treaty. And there exists an even stronger positive correlation than the one noted above for the International Fact-finding Commission between acceptance of this humanitarian treaty and a state's levels of wealth, industrialization, human development, political freedom, and level of governmental integrity (non-corruption).

8.3.2 Intrastate Warfare

1949 Geneva Convention IV, although applicable almost entirely to interstate warfare, does provide (via its Article 3 [an article common to all four of this set of Geneva Conventions]) for some minimal social protections in the case of 'armed conflict not of an international character'. However, the only major humanitarian instrument devoted to intrastate warfare is 1977 Protocol II on Non-international Armed Conflicts, a companion to the 1977 Protocol I already discussed. 1977 Protocol II is rather mild in its strictures, presumably owing to the potential for rejection on the basis of undermining national sovereignty and for the fear that it might legitimize, and perhaps even encourage, insurgencies (Lopez, 1994). As to present concerns, its only matters of special environmental interest are—in parallel with 1977 Protocol I—certain constraints on the destruction of agricultural areas (Article 14), and also on the release of dangerous forces (Article 15). Regrettably, there is nothing in this treaty akin to the provisions in its companion instrument for sparing the natural environment as such, or for an optional International Fact-finding Commission. More than two-thirds of all states have so far become a party to this treaty. As was the case with 1977 Protocol I, states have joined essentially irrespective of their national wealth, industrialization, human development, or political status (cf. Table 8.1). And again, those states with past experience in major warfare in recent decades have been less apt to become parties. (Of the 11 states parties to Protocol I that have to date chosen not to join Protocol II, about three-quarters are politically oppressed.)

8.4 Treaty Compliance

Issues of compliance—implementation, monitoring, verification, and enforcement—of multilateral treaties (whether International Environmental Law, International Humanitarian Law, or other) often are politically cumbersome and sensitive concerns. Although the effectiveness of international environmental treaties is of great importance to the future of the global biosphere and human well-being (French, 1994; Mitchell, 1994; Sand, 1992; Sands, 1993), compliance issues associated with this body of international law do not seem readily applicable to the concerns being addressed here. The following discussion thus draws primarily on information gained from International Humanitarian Law.

As shall be explained below, it does not seem possible at this point to offer direct evidence of compliance by states with the three explicitly environmentally pertinent contributions to International Humanitarian Law that have been singled out for discussion. Nonetheless, it can be proposed as a working hypothesis that future compliance by states parties to the environmental provisions of those treaties might, in fact, be reasonably satisfactory. This is being suggested for at least three reasons (Chayes & Chayes, 1993; 1995):

(a) *Analysis of self-interest:* The first factor that supports treaty compliance is that states—especially the major and reasonably major powers—do not become party to an International Humanitarian Law (or other) multilateral treaty without prior detailed governmental review of the pros and cons of such a commitment. They carry out such a review from a panoply of political, military, economic, and other considerations, a process that might well take many months or even years to complete. If it is an incipient treaty still under negotiation, a potential state party will, of course, attempt to revise the instrument at that stage to something palatable to it. In the end, a treaty under consideration will simply be rejected by a state unless its advantages are clearly seen to outweigh its disadvantages. For example, the 1980 Land Mine Protocol has so few states parties for the simple reason that land mines are perceived by many powers to be too militarily valuable (and cost effective) to give up for humanitarian reasons (but cf. below). And it has already been noted that the nuclear powers have to date refused to relinquish the to them politically and militarily attractive, if not indispensable, option to wage nuclear war.

(b) *Underlying cultural norms:* A second factor that supports treaty compliance is that the provisions of an International Humanitarian Law treaty do not exist in a vacuum, but express legal norms that have in turn derived from, and rest upon, germane cultural norms (Westing, 1988a; 1988c). That is to say, unless a treaty in this body of law reflects general societal attitudes, it will not be subscribed to in the first place. Thus, the likelihood of non-compliance—certainly of flagrant non-compliance—by a state party would be low. Of course, occasional breaches cannot be ruled out, comparable, say, to domestic transgressions of the clear cultural and legal norms against murder. Take the case of the cultural norm against the use in war of chemical or biological weapons, and its cardinal legal expression via the 1925 Geneva Protocol. This treaty—an international response to the horrors of the extensive chemical warfare carried out by both sides during World War I—has been respected by most of the states parties as regards both of its proscriptions (chemical and biological) since its entry into force in 1928 (Westing, 1985b; 1988b). Moreover, it is important to point out here that the rare violations of either proscription (whether by states parties or non-parties) have been perpetrated under conditions of utmost secrecy, a demonstration of the strength of the underlying cultural norm against such uses. As to the case of land mines, growing awareness of their horrible postwar social impact—the growing awareness and relevant cultural norms developing in significant part through the intense efforts of a number of nongovernmental organizations—may in time lead to the widespread adoption of stronger legal norms. Indeed, one of the more interesting recent developments in the international system is the rise in numbers and importance of nongovernmental organizations dealing with environmental, human rights, and humanitarian concerns, this development tending to undermine somewhat the governmental monopoly in international relations (Westing, 1994b, p. 214).[3] Finally it must be noted that a treaty, once entered into, serves—through its concrete articulation and authoritative dissemination—to serve both as a confidence-building measure (Goldblat, 1994, pp 209–210) and to reinforce the cultural norms on which it was based (Stone, 1988). This latter normative role of a treaty seems to be especially pronounced in the case of International Human Rights Law (Chayes & Chayes, 1993, p. 197).

(c) *Treaty ambiguity:* A third factor that supports treaty compliance is the often wide range of interpretations applicable to the agreed upon constraints. How in a practical sense is the following otherwise undefined basic rule in 1977 Protocol I to actually be interpreted and complied with by a military commander in the field: 'It is prohibited to employ methods or means of warfare which are intended, or may be expected, to cause widespread, long-term and severe damage to the natural environment' (Article 35.3; cf. Article 55.1)? What about the required precaution to refrain from 'any attack which may be expected to cause...damage to civilian objects...which would be excessive in relation to the concrete and direct military advantage anticipated' (Article

[3] Two treaties of relevance here have, in fact, entered into force following publication of this paper, both in large part owing to nongovernmental pressures: (**a**) the 1997 Anti-personnel Land Mine Convention (*UNTS* 35597); and (**b**) the 2008 Cluster Munition Convention (*UNTS* 47663).

57.2.iii)? And how is the prohibition against releasing certain dangerous forces to be interpreted by the field commander in the light of the qualifying phrase, 'if such attack may cause…consequent severe losses among the civilian population' (Article 56.1)? It becomes clear from these ambiguous examples of limitation (and various other similar ones that could be cited) that the wide latitudes involved, and subjective evaluations required, in complying with them make the nurturing of the relevant underlying cultural norms an absolutely crucial component of meaningful compliance.

8.4.1 Verification

The monitoring or verification machinery associated especially with nuclear-weapon and chemical-weapon international arms control treaties can become truly elaborate and expensive (Goldblat, 1994, pp 209–243; Krass, 1985; 1996). As to interstate warfare in general, one available verification procedure of special interest in the present context has already been touched upon earlier, namely the optional International Fact-finding Commission associated with 1977 Protocol I and the four 1949 Geneva Conventions. The low level—and politically, economically, and developmentally skewed distribution—of this straightforward mechanism of verification must be attributed to a variety of factors, among them notions of infringement of national sovereignty, the associated administrative and financial burdens, and the lack of a thoroughly serious commitment to the legal norms. As to intrastate warfare, the International Fact-finding Commission is not applicable, nor is there any other comparable mechanism in place. It is regrettable that more states do not submit to such verification mechanisms because such submission would act as a further deterrent to any contemplated breaches.

The ultimate forum available to a state to attempt to verify (establish) non-compliance with a treaty by a second state is the International Court of Justice (The Hague), open to all members of the United Nations (and to other states under special conditions). However, accepting the competence of the Court on a compulsory and unconditional basis is optional to the states parties to the 1945 United Nations Charter (cf. Court Statute, Article 36). The fervor with which notions of national sovereignty are embraced by states—that is, the actual degree of international anarchy—is exemplified by the fact that only two dozen or so of them have accepted the jurisdiction of the Court on such a basis (Westing, 1990).

8.4.2 Enforcement

The matter of enforcing environmental protections from wartime damage that have been established by international law can take several forms, for various reasons none of them fully satisfactory. During the course of a war, the threat of reprisal can be used to protect the environment from damage; and actual reprisal attacks can be carried out after the event. As to interstate warfare, attacks in reprisal are permitted by International Humanitarian Law. However, in the event of hostile environmental damage covered by 1977 Protocol I, such attacks in reprisal are not permitted to be carried out in kind—more specifically, not against the natural environment (Article 55.2), not against certain agricultural areas (Article 54.4), and not via the release of certain dangerous forces (Article 56.4). As to intrastate warfare, International Humanitarian Law is essentially silent on the issue of enforcement, 1977 Protocol II, for example, not even mentioning the notion of reprisal.

Under some conditions, the mobilization of public opinion, both domestic and international, can be a valuable approach to enforcement. Certainly the widespread revulsion to environmental destruction by the USA during the Second Indochina War ('Vietnam Conflict') of 1961–1975 contributed to its ultimate cessation (and, as noted below, to the subsequent enactment of certain treaty constraints). As a further approach, special tribunals, under United Nations or other auspices, can be established to examine and judge breaches and to impose punishment, although it must be noted that damage to the natural environment, as prohibited by 1977 Protocol I, has not as such been designated as a grave breach, that is, a war crime (cf. Article 85). On the other hand, the plundering of forests by Germany in occupied Poland during World War II was classed as a war crime by the United Nations War Crimes Commission at Nuremberg (its Case No. 7150), presumably as a violation of Annex Article 55 of 1907 Hague Convention IV (UNWCC, 1948, p. 496). The current efforts within the United Nations to establish a permanent international criminal court to try alleged violations of International Humanitarian Law are most commendable (*UNGA*, 1995a).[4] Of course, complaints by a member state over any breach of the peace can be brought before the *United Nations Security Council* (which can also act on its own initiative), with the 1945 United Nations Charter providing for means to institute coercive collective sanctions (Article 41) and even military actions (Article 42); a relevant Security Council action is referred to below in relation to the Persian Gulf War of 1991.

8.4.3 Two Specific Wars

Two wars of recent decades that have been especially disruptive of the environment—the Second Indochina War of 1961–1975 and the Persian Gulf War of 1991—do not lend themselves to testing the efficacy of the three outlined treaties having specific environmental relevance. Nonetheless, it is necessary to examine those two wars at least briefly, both separately and collectively, from the

[4] The 1998 International Criminal Court Statute (*UNTS* 38544) entered into force following publication of this paper.

overall standpoint of compliance with International Humanitarian Law in relation to environmental disruption:

(a) *Second Indochina War of 1961–1975:* In the Second Indochina War, the USA brought about immense amounts of environmental damage over the course of some years in Cambodia, Laos, and especially Viet Nam by destroying huge forest and agricultural areas through various chemical, explosive, and mechanical means (Westing, 1976; 1989a). The chemical attacks contravened the chemical prohibition in the 1925 Geneva Protocol referred to earlier, but is not applicable to this war since none of the four relevant states were party to that treaty at the time (all but Cambodia now are). 1977 Protocol I—as well as another treaty of only marginal interest, the 1977 Environmental Modification Convention—would have been relevant, but did not exist at the time. Indeed, the environmental protections introduced by those two instruments were an international response to that war. The 1980 Land Mine Protocol also did not exist at the time (and only Laos is now a party). And no specific restraints exist on land clearing via massive bombing or the use of tractors. More generalized (non-specific) restraints of relevance are touched upon below.
(b) *Persian Gulf War of 1991:* In the Persian Gulf War, the USA and other United Nations Coalition Forces over the course of some weeks caused environmental damage through their extraordinarily intensive bombing campaign, but it was the massive releases by Iraq of Kuwaiti oil that have made this war notorious from an environmental standpoint (Westing, 1994a). The release of oil in liquid form befouled large areas of Kuwait's terrestrial environment as well as inshore portions of the Persian Gulf. Further huge amounts of escaping oil were set on fire, thereby saturating the local atmosphere with dense, noxious smoke for a period of months. The protection of the natural environment deriving from 1977 Protocol I would have been relevant, but although Kuwait was a party, neither Iraq nor the USA were (or are). What protection the 1980 Land Mine Protocol might have afforded was also not applicable since none of the three relevant states were (or are) parties to that treaty. Again, more generalized (non-specific) restraints of relevance are touched upon below.
(c) *The two wars:* It has been seen that the practical efficacy of the three treaties in question of special environmental interest cannot be tested on the basis of the two environmentally notorious wars just presented. However, the environmental destruction in both of those wars was presumably covered, by virtue of being 'enemy property', by 1899 Hague Convention II and 1907 Hague Convention IV as well as by 1949 Geneva Convention IV, which, *inter alia*, proscribe the destruction of enemy property during hostilities, unless it be imperatively demanded by the necessities of war (both Hagues, Annex Article 23.g), or during occupation (both Hagues, Annex Article 55; Geneva, Article 53). Of the six main states involved in the two wars, only the USA was a state party to the two Hague Conventions (and none of the others is as yet), but it is generally accepted that this pair of treaties has achieved a status of 'customary' international law, that is, to be unavoidably binding upon all nations. Geneva

Convention IV is also applicable to both of the wars since all of the six main states involved in them were states parties during the relevant periods (and, in any case, it is also generally considered to have achieved 'customary' status). Regarding the Persian Gulf War, it is thus important to note that the *United Nations Security Council* resolved that Iraq was liable for any direct environmental damage and depletion of natural resources caused by its invasion and occupation of Kuwait (UNSC, 1991b, ¶ 16), presumably on the basis of Article 3 of 1907 Hague Convention IV (cf. also 1977 Protocol I, Article 91). The Security Council further demanded of Iraq (apparently without specific legal basis) that it supply information on the locations of the land mines and similar devices it had emplaced in Kuwait, so as to facilitate their clearance (UNSC, 1991a, ¶ 3.d). As an aside, it is of interest to note here that what may have been the first formal recognition by the Security Council of wartime environmental damage occurred during the Iran–Iraq War of 1980–1988, in which it called upon both parties to refrain from any action that could endanger marine life in the region of the Persian Gulf (UNSC, 1983, ¶ 5).

8.5 Implications

Given that warfare—a recurring human pursuit—is by its very nature not only a barbaric and lethal pastime, but also an environmentally disruptive one, to what extent can such environmental damage be ameliorated by international law? International Humanitarian Law (the Law of War), including International Arms Control and Disarmament Law, has developed over the decades as a response to the deadly and malevolent nature of warfare, attempting primarily to minimize particularly cruel and perfidious interstate military actions and to protect wounded combatants, prisoners of war, and enmeshed civilians. Some constraints embedded within International Humanitarian Law are incidentally protective of the environment. However, with the growing recognition of the central importance of an increasingly abused environment for human survival and well-being, the scope of this body of law has in recent years been expanded to specifically embrace some aspects of environmental protection. One encouraging manifestation of such recognition is the support recently given by a great majority of states to incorporating environmental considerations into the negotiation and implementation of arms control and disarmament treaties (*UNGA*, 1995b, ¶ 1).

It is clearly not an easy matter to establish the extent to which International Humanitarian Law is actually conducive to ameliorating the social and environmental horrors of war. As one bit of anecdotal evidence supporting the efficacy of such treaties, an official of the *International Committee of the Red Cross* (Geneva) found that during World War II, Germany—a state party to the 1929 Geneva Convention Relative to the Treatment of Prisoners of War (*LNTS* 2734)—did, on the basis of reciprocity, treat its prisoners of war from other states parties (*e.g.*, of the United Kingdom) better than from those of states non-parties (*e.g.*, of Russia)

(Junod, 1951, pp 227–229, etc.). But to what extent was the noted correlation the result of treaty membership patterns, and to what extent had prisoner treatment been the result of other factors? However, the earlier discussion of the process of treaty adoption, together with the comparison made above between states parties and non-parties to certain International Humanitarian Law treaties, certainly suggest that treaty membership is determined primarily by state positions (and their underlying cultural norms).

The question arises of whether existing constraints on environmental damage by warfare suffice or whether they should be augmented in one way or another. A major argument for strengthening the provisions is the ambiguities or other weaknesses involved in the current ones; another is that the very process of negotiating more restrictive provisions serves an educational (normative) function (Falk, 1992; Verwey, 1995). One of the more valuable strengthening initiatives has been the drafting by a private party of a proposed 'Convention on the Crime of Ecocide' (Falk, 1984). A more circumscribed initiative of that sort has been the proposed 'Treaty for Protection Against Nuclear Devastation' (Westing, 1989b).

A major argument for attempting no significant additions to the existing constraints on environmental damage by warfare is that the current constraints are already at the limits of current acceptability—or even a bit beyond them (Bouvier, 1991; 1992; Goldblat, 1991; Roach, 1997; Roberts, 1992; Westing, 1994b). In the case of interstate warfare this is evidenced by the low level of acceptance of some of the pertinent treaties or of their important optional provisions (*e.g.*, the 1980 Land Mine Protocol and Article 90 of 1977 Protocol I). In the case of intrastate warfare, this is evidenced by the paucity of relevant treaty law, and the substantial weaknesses of the single existing one (1977 Protocol II). It is thus presumably correctly argued that what is necessary at this point is to strive for: (a) more nearly universal formal adherence to the existing body of International Humanitarian Law; (b) more thorough dissemination of this body of law, and of the incorporation of its provisions into the military manuals used by the armed forces of the world (*ICRC*, 1993); and, above all, (c) increased relevant discussion and education, both formal and informal, in both the civil and military sectors of society (Westing, 1993b). The growing awareness of environmental concerns within the military sector is a welcome trend (*e.g.*, Diederich, 1992; Drucker, 1989).

The comparisons presented earlier between states parties and non-parties to a number of environmentally relevant instruments (cf. Table 8.1) additionally suggest the need for a broader long-term agenda than just outlined if the goal of universal acceptance of International Humanitarian Law is, in fact, to be approached. The data suggest that it will be necessary for democratization on the one hand and overall development on the other to be much more widely prevalent among the many states of the world to achieve the desired goal. In each of the several indicators examined there is evidence of a clear gradient of increasing acceptance by states of the key instruments as the states' conditions improve. To illustrate, those states considered to be free (*i.e.*, enjoying high levels of political rights and civil liberties) are three times as apt to have opened their borders to an International Fact-finding Commission for the investigation of alleged infractions of 1977 Protocol I or of the four 1949 Geneva Conventions than those states at the low end of the freedom spectrum. Similarly, the free states are almost four times as apt to have accepted the strictures imposed by the 1980 Land Mine Protocol. The wealthiest states (in terms of gross national product per caput) are five times as apt to have accepted that Fact-finding Commission than the poorest states; and more than three times as apt to have accepted land mine strictures. Such comparisons become even more stark when made in terms of a state's level of industrialization or of its human development (the latter as measured in terms of life expectancy, adult literacy, and wealth).

As an aside, it is interesting to note that the interstate wars in which democratic regimes engage are apt to lead to fewer fatalities than those of non-democratic regimes (Rummel, 1995), perhaps a reflection of greater adherence to International Humanitarian Law. It might additionally be suggested here that a spreading of democracy and overall development among states will provide the added bonus of reducing the frequency of intrastate (internal) wars, which are so intractable in terms of International Humanitarian Law.

The earlier argument for maintaining the limitation *status quo* notwithstanding, two strengthenings of existing constraints on environmental damage by warfare must be recommended at this time. The first is a weapon-specific treaty that would—in view of their devastating impacts on both humans and nature—prohibit the use in war of nuclear weapons (Westing, 1989b). The second is that at least the more important legally established nature reserves—especially those designated by the 1972 World Heritage Convention as natural heritage sites of outstanding universal value—be formally designated as demilitarized zones in accordance with Article 60 of 1977 Protocol I (Antoine, 1992; Westing, 1992b). In can almost go without saying that protecting such reserves from depredations, both civil and military, is crucial to the conservation of the global biosphere, for purposes of maintaining essential ecosystem functioning, and of protecting biodiversity.

In closing, the role of international law in protecting the environment from military damage is of clear importance because it articulates rather precisely the state of the cultural norms upon which it rests, because it reinforces those norms somewhat by its very existence, and because it elucidates the shortcomings in those norms which require progressive development through education and otherwise.

References

Antoine, P. 1992. International Humanitarian Law and the protection of the environment in time of armed conflict. *International Review of the Red Cross* (Geneva) 32(291):517–537.

Bouvier, A. 1991. Protection of the natural environment in time of armed conflict. *International Review of the Red Cross* (Geneva) 25(285):567–578.

Bouvier, A. 1992. Recent studies on the protection of the environment in time of armed conflict. *International Review of the Red Cross* (Geneva) 32(291):554–566.

Chayes, A., & Chayes, A.H. 1993. On compliance. *International Organization* (Cambridge, MA, USA) 47(2):175–205.

Chayes, A., & Chayes, A.H. 1995. *The New Sovereignty: Compliance with International Regulatory Agreements*. Cambridge, MA, USA: Harvard University Press, 417 pp.

CIA. 1995. *World Factbook 1995*. Washington: US Central Intelligence Agency, 557 pp + 17 maps.

Diederich, M.D., Jr. 1992. "Law of War" and ecology: a proposal for a workable approach to protecting the environment through the Law of War. *Military Law Review* (Charlottesville, VA, USA) 136:137–160.

Drucker, M.P. 1989. Military commander's responsibility for the environment. *Environmental Ethics* (Denton, TX, USA) 11:135–152.

Falk, R.A. 1975. Methods and means of warfare: counterinsurgency, tactics, and the law. In: Trooboff, P.D. (ed.). *Law and Responsibility in Warfare: the Vietnam Experience*. Chapel Hill, NC, USA: University of North Carolina Press, 280 pp: pp 35–53,102–113.

Falk, R.A. 1984. Proposed Convention on the Crime of Ecocide. In: Westing, A.H. (ed.). *Environmental Warfare: a Technical, Legal and Policy Appraisal*. London: Taylor & Francis, 107 pp: pp 45–49.

Falk, R.[A.]. 1992. Environmental Law of War: an Introduction. In: Plant, G. (ed.). *Environmental Protection and the Law of War: a 'Fifth Geneva' Convention on the Protection of the Environment in Time of Armed Conflict*. London: Belhaven Press, 284 pp: pp 78–95.

French, H.F. 1994. Making environmental treaties work. *Scientific American* (New York) 271(6):94–97.

Goldblat, J. 1991. Legal protection of the environment against the effects of military activities. *Bulletin of Peace Proposals* [now *Security Dialogue*] (Oslo) 22(4):399–406.

Goldblat, J. 1994. *Arms Control: a Guide to Negotiations and Agreements*. London: Sage Publications, 772 pp.

IBRD. 1995. *World Development Report 1995: Workers in an Integrating World*. Washington: *International Bank for Reconstruction & Development [World Bank]* -&- New York: Oxford University Press, 251 pp.

ICRC. 1993. *Guidelines for Military Manuals and Instructions on the Protection of the Environment in Time of Armed Conflict*. Geneva: *International Committee of the Red Cross*, 6 pp; also in: New York: *United Nations General Assembly*, Document No. A/49/323 (19 Aug 94), 53 pp: pp 49–53; also in: *American Journal of International Law* (Washington) 89(3):641–644. 1995.

Junod, M. 1951. *Warrior without Weapons* [transl. from the French by Fitzgerald, E.]. London: Jonathan Cape, 318 pp.

Kaplan, R. (ed.). 1996. *Freedom in the World: the Annual Survey of Political Rights & Civil Liberties 1995–1996*. New York: Freedom House, 545 pp.

Krass, A.S. 1985. *Verification: How Much is Enough*? London: Taylor & Francis, 271 pp.

Krass, A.S. 1996. *Costs, Risks, and Benefits of Arms Control*. Stanford, CA, USA: Stanford University, Center for International Security & Arms Control, 51 pp.

Lambsdorff, J.G. 1996. *1996 Transparency International Corruption Perception Index: an Index of Perceptions of Business People of Corruption around the World*. Berlin: Transparency International, 8 pp.

Lopez, L. 1994. Uncivil wars: the challenge of applying International Humanitarian Law to internal armed conflicts. *New York University Law Review* (New York) 69(4–5):916–962.

Mitchell, R.B. 1994. Regime design matters: intentional oil pollution and treaty compliance. *International Organization* (Cambridge, MA, USA) 48(3):425–458.

Roach, J.A. 1997. The Laws of War and the protection of the environment. *Environment & Security* (Ste-Foy, Canada) 1(2):53–67.

Roberts, A. 1992. Environmental destruction in the 1991 Gulf War. *International Review of the Red Cross* (Geneva) 32(291):538–553.

Rummel, R.J. 1995. Democracies ARE less warlike than other regimes. *European Journal of International Relations*, (London) 1(4):457–479.

Sand, P.H. 1992. *Effectiveness of International Environmental Agreements: a Survey of Existing International Instruments*. Cambridge, UK: Grotius Publications, 539 pp.

Sands, P. 1993. Enforcing environmental security: the challenges of compliance with international obligations. *Journal of International Affairs* (New York) 46(2):367–390.

Sivard, R.L. 1996. *World Military and Social Expenditures 1996*. 16th edn. Washington: World Priorities, 56 pp.

Stone, C.D. 1988. The law as a force in shaping cultural norms relating to war and the environment. In: Westing, A.H. (ed.). *Cultural Norms, War and the Environment*. Oxford: Oxford University Press, 177 pp: pp 64–82.

UNDP. 1995. *Human Development Report 1995*. 6th edn. New York: United Nations Development Programme -&- New York: Oxford University Press, 230 pp.

UNGA. 1995a. *Establishment of an International Criminal Court*. New York: *United Nations General Assembly*, Resolution No. 50/46 (11 Dec 95), 2 pp. [185 (100 %) in favor (adopted without a vote)]

UNGA. 1995b. *General and Complete Disarmament: Observance of Environmental Norms in the Drafting and Implementation of Agreements on Disarmament and Arms Control*. New York: *United Nations General Assembly*, Resolution No. 50/70 M (12 Dec 95), 1 p. [157 (85 %) in favor, 2 abstentions, 4 against, 22 absent = 185]

UNSC. 1983. *[Situation between Iran and Iraq.]* New York: *United Nations Security Council*, Resolution No. S/RES/540(1983) (31 Oct 83), 1 p. [12 (80 %) in favor, 3 abstentions, 0 against, 0 absent = 15]

UNSC. 1991a. *[Suspension of offensive combat operations.]* New York: *United Nations Security Council*, Resolution No. S/RES/686(1991) (2 Mar 91), 3 pp. [11 (73 %) in favor, 3 abstentions, 1 against, 0 absent = 15]

UNSC. 1991b. *[The restoration to Kuwait of its sovereignty.]* New York: *United Nations Security Council*, Resolution No. S/RES/687(1991) (3 Apr 91), 10 pp. [12 (80 %) in favor, 2 abstentions, 1 against, 0 absent = 15]

UNWCC. 1948. *History of the United Nations War Crimes Commission and the Development of the Laws of War*. London: His Majesty's Stationery Office, 592 pp.

USDOS. 1993. *Hidden Killers: the Global Problem with Uncleared Landmines*. Washington: US Department of State, Publication No. 10098, 185 + 84 pp.

USDOS. 1994. *Global Landmine Crisis*. Washington: US Department of State, Publication No. 10225, 61 + 25 pp.

Verwey, Wil D. 1995. Protection of the environment in times of armed conflict: in search of legal perspectives. *Leiden Journal of International Law* (Leiden) 8(1):7–40.

Westing, A.H. 1976. *Ecological Consequences of the Second Indochina War*. Stockholm: Almqvist & Wiksell, 119 pp + 8 pl.

Westing, A.H. 1982. War as a human endeavor: the high-fatality wars of the twentieth century. *Journal of Peace Research* (Oslo) 19(3):261–270.

Westing, A.H. 1984. Environmental warfare: an overview. In: Westing, A.H. (ed.). *Environmental Warfare: a Technical, Legal and Policy Appraisal*. London: Taylor & Francis, 107 pp: pp 3–12.

Westing, A.H. 1985a. Explosive remnants of war: an overview. In: Westing, A.H. (ed). *Explosive Remnants of War: Mitigating the Environmental Effects*. London: Taylor & Francis, 141 pp: pp 1–16.

Westing, A.H. 1985b. Towards eliminating the scourge of chemical war: reflections on the occasion of the sixtieth anniversary of the Geneva Protocol. *Bulletin of Peace Proposals* [now *Security Dialogue*] (Oslo) 16(2):117–120.

Westing, A.H. 1987. Ecological dimension of nuclear war. *Environmental Conservation*. (Cambridge, UK) 14(4):295–306.

Westing, A.H. 1988a. Constraints on military disruption of the biosphere: an overview. In: Westing, A.H. (ed.). *Cultural Norms, War and the Environment*; Oxford: Oxford University Press, 177 pp: pp 1–17.

Westing, A.H. 1988b. Cultural constraints on warfare: micro-organisms as weapons. *Medicine & War* (London) 4(2):85–95.

Westing, A.H. 1988c. Towards non-violent conflict resolution and environmental protection: a synthesis. In: Westing, A.H. (ed.). *Cultural Norms, War and the Environment*. Oxford: Oxford University Press, 177 pp: pp 151–159.

Westing, A.H. 1989a. Herbicides in warfare: the case of Indochina. In: Bourdeau, P., *et al.* (eds). *Ecotoxicology and Climate: with Special Reference to Hot and Cold Climates*. Chichester, UK: John Wiley, 392 pp: pp 337–357.

Westing, A.H. 1989b. Proposal for an International Treaty for Protection Against Nuclear Devastation. *Bulletin of Peace Proposals* [now *Security Dialogue*] (Oslo) 20(4):435–436.

Westing, A.H. 1990. Towards eliminating war as an instrument of foreign policy. *Bulletin of Peace Proposals* [now *Security Dialogue*] (Oslo) 21(1):29–35.

Westing, A.H. 1992a. Environmental dimensions of maritime security. In: Goldblat, J. (ed.). *Maritime Security: the Building of Confidence*. Geneva: United Nations Institute for Disarmament Research, Document No. UNIDIR/92/89, 159 pp: pp 91–102.

Westing, A.H. 1992b. Protected natural areas and the military. *Environmental Conservation* (Cambridge, UK) 19(4):343–348.

Westing, A.H. 1993a. Environmental Modification Convention: 1977 to the present. In: Burns, R.D. (ed.). *Encyclopedia of Arms Control and Disarmament*. New York: Charles Scribner's Sons, 1692 pp: pp 947–954.

Westing, A.H. 1993b. Global need for environmental education. *Environment* (Washington) 35(7):4–5,45.

Westing, A.H. 1993c. Human rights and the environment. *Environmental Conservation* (Cambridge, UK) 20(2): 99–100.

Westing, A.H. 1994a. Constraints on environmental disruption during the Gulf War. In: O'Loughlin, J., *et al.* (eds). *War and its Consequences: Lessons from the Persian Gulf Conflict*. New York: HarperCollins, 252 pp: pp 77–84.

Westing, A.H. 1994b. Environmental change and the international system: an overview. In: Calließ, J. (ed.). *Treiben Unweltprobleme in Gewaltkonflikte?: Ökologische Konflikte im internationalen System und Wege zur Kooperation*. Rehburg-Loccum, Germany: Evangelische Akademie Loccum, Loccumer Protokolle, No. 21/94, 354 pp: pp 207–228.

Chapter 9
Protecting the Environment in War: Military Guidelines

Note: *A state's (nation's) military manuals and rules of engagement (often publicly available) serve to guide the armed forces of that state during both their peacetime and wartime actions. The constraints contained in those documents derive, of course, to a greater or lesser degree from the existing Law of War (International Humanitarian Law and related International Arms Control and Disarmament Law), as outlined earlier* (cf. Chap. 8). *Interestingly enough, in various instances a state may lean on the Law of War even if it is not a state party to one or more of the relevant multilateral instruments. Moreover, some states make use (and more should) of the relevant model guidelines for such documents that have been developed by the International Committee of the Red Cross in Geneva. What is reproduced below (**#357**)*[1] *touches upon peacetime and wartime guidelines and also offers some suggestions for the future.*[2]

9.1 Background

Unsustainable discharges of waste gases into the atmosphere and large numbers of species extinctions throughout the world are but two of many indications of the increasingly deleterious impact by humankind on the global biosphere. With the

[1] The numbered references are provided in Chap. 3.

[2] Reproduced from: Austin, J.E., & Bruch, C.E. (eds). *The Environmental Consequences of War: Legal, Economic, and Scientific Perspectives*. Cambridge, UK: Cambridge University Press, 691 pp: pp 171–181 (Chap. 6); 2000 with the original title *"In Furtherance of Environmental Guidelines for Armed Forces during Peace and War"* by permission of the Environmental Law Institute, the copyright holder, on 22 March 2012. Invited paper, 'First International Conference on Addressing Environmental Consequences of War: Legal, Economic, and Scientific Perspectives', Washington, 10–12 June 1998, of the Environmental Law Institute *et al.* The author is pleased to acknowledge information from Carl Bruch (Washington), Jean-Marie Henckaerts (Geneva), and Masa Nagai (Nairobi); and suggestions from Richard C. Tarasofsky (Bonn) and Carol E. Westing (Putney).

A. H. Westing, *Arthur H. Westing*, SpringerBriefs on Pioneers in Science and Practice 1, DOI: 10.1007/978-3-642-31322-6_9,

civil sector of society responsible for most of this abuse, it is only natural that attempts at ameliorative action are directed almost exclusively toward that sector. However, a number of arguments can readily be mustered to suggest the importance of not overlooking the military sector of society in conserving the environment.

First, although military activities now contribute only about 3 % to total human activities worldwide (as measured in terms of gross national products) (*ACDA*, 1997, p. 49), every bit of ameliorative action is valuable in the increasingly dire environmental circumstances prevailing today. *Second*, some military activities have the potential for being environmentally disruptive at levels disproportionately high in relation to their contribution to overall human activities, thus requiring particular attention—recall here, among others: the Yellow River valley/1938; Gruinard Island/1942; northern Norway/1944; Hiroshima and Nagasaki/1945; Enewetak (Eniwetok) atoll/1952; Viet Nam and Laos/1970; Kuwait and the Persian Gulf/1991; Eritrea/1991; Estonia and Latvia/1991; and Cambodia/1992 (Westing, 1980; etc.). *Third*, there is a tendency for the military sector to consider itself immune from applicable restraints on environmental abuse, especially so during wartime, but also during peacetime. *Fourth*—some would add—the military sector to some greater or lesser extent does not contribute to human welfare and thus becomes a prime candidate for curtailment.

Since there are many who are unaware of—or perhaps unwilling to accept—the pervasiveness of military activities in the world, it will be useful to be reminded that some 163 of the 192 sovereign states into which we have sorted ourselves maintain regular armed forces (*ACDA*, 1997, p. 36); and, indeed, that about 10 % of all government expenditures in the world today are devoted directly to maintaining those regular forces (*ACDA*, 1997, p. 49). Additionally, there always exist 30 or more insurgent forces, although any one of them on a somewhat less permanent basis. And, although it is true that many states are at peace much of the time—with their armed forces engaged primarily in training, garrison duty, patrolling, weapon testing, and serving as a threat—from time to time they do also engage in combat, both beyond and (now more frequently) within their own borders. Indeed, well over a hundred governments have made hostile use of their armed forces merely since the end of World War II in support of their multifarious foreign and domestic policy agendas (Sivard, 1996, pp 18–19; Smith, 1997, pp 90–95; Tillema, 1989; Westing, 1982).

The extent to which societal concerns over the deteriorating global environment have extended into military sectors is described below, primarily as a means of supporting them and facilitating their spread. The peacetime situation is noted first, followed by that of wartime. Touched upon finally are some thoughts of where we must go from here, primarily at this stage in order to reveal means for achieving a wider acceptance of the existing military guidelines.

9.2 Peacetime Guidelines

With the military sector of a state widely accepted to be concerned with supreme national interests, it is equally widely taken for granted that the military sector is beyond the reach of a state's civil sector, both in democratic and totalitarian states. Indeed, in at least four states—Germany, Serbia/Montenegro, Switzerland, and the United Kingdom—the armed forces are explicitly exempted in whole or in part from domestic environmental protection legislation (UNEP, 1995b; 1995c; 1995d). Moreover, numerous multilateral environmental protection treaties dealing with the marine environment specifically exempt naval ships from their constraints (Westing, 1992a). And, at US insistence, the 1997 Kyoto Protocol to the 1992 Framework Convention on Climate Change (*UNTS* 30822) also includes a military exemption provision (Warrick, 1998).

It is thus gratifying to point out that in at least 19 states—Bangladesh, Croatia, Denmark, Finland, India, Indonesia, Iran, Malaysia, Maldives, Netherlands, Norway, Pakistan, Poland, South Africa, Sri Lanka, Sweden, Thailand, Viet Nam, and USA—national environmental protection legislation applies equally to the military and civil sectors, at least domestically during peacetime (UNEP, 1995b; 1995d; 1996a; 1996b). Moreover, the *North Atlantic Treaty Organization* (NATO) (Brussels) has recently developed a set of quite detailed environmental guidelines for armed forces during peacetime, going on to suggest that these would be appropriate for any state to adopt (NATO, 1996). The NATO guidelines promote environmental responsibility and in essence urge that, within limits, the military sector of a state should comply with the environmental rules established for its civil sector. Indeed, through its own sound environmental practices the military sector should, as *NATO* would have it, be serving as an example to the rest of society.

Even in the absence of military/civil parity before domestic law, it is clear that environmental concerns are beginning to pervade the armed forces of the world. The defense ministries of at least the following 11 states have in recent years established permanent environmental divisions and programs: Bulgaria, Croatia, Czech Republic, Denmark, Germany, Hungary, Pakistan, Sweden, United Kingdom, USA, and Viet Nam (*UNEP*, 1995b; 1995c; 1995d; 1996a; 1996b). The USA appears to have done this more thoroughly and elaborately than any other state (Goodman, 1994; 1997; Renew America, 1995; USDoD, 1996), with one high Pentagon official proudly referring to the US armed forces as now being 'lean, mean, and green'. Moreover, for better or worse, 13 or more states assign to their armed forces the enforcement of environmental protection laws, including: Bangladesh, Bhutan, Cambodia, India, Indonesia, Laos, Malaysia, Maldives, Myanmar, Nepal, Philippines, Sri Lanka, and Thailand (UNEP, 1996a; 1996b).

Before leaving peacetime environmental guidelines for the military sector it is useful to note the role of the United Nations in sensitizing the world at large to these issues. Prompted in part by the Programme of Action for Sustainable

Development ('Agenda 21') adopted by the 1992 United Nations Conference on Environment and Development (UN, 1993, p. 201, ¶ 20.22[h]), the *United Nations Environment Programme* (UNEP) (Nairobi) in 1993 sent an appropriate questionnaire to all states (UNEP, 1993; 1995b). Then, on the advice of the *United Nations Commission on Sustainable Development* (UNCSD, 1994, chap. 1, ¶¶ 186–187), in 1995 and 1996 followed up its questionnaire with a series of three regional intergovernmental conferences (UNEP, 1995a; 1995c; 1995d; 1996a; 1996b). And the *United Nations General Assembly* has urged that all states observe environmental norms in the drafting and implementation of additions to International Arms Control and Disarmament Law (UNGA, 1995; 1996; 1997).

9.3 Wartime Guidelines

Any consideration of constraints on environmental disruption during wartime must perforce distinguish between international armed conflicts and the now far more common non-international (internal) armed conflicts. This distinction is important because the Law of War (International Humanitarian Law) is applicable primarily to international armed conflicts. Thus the Law of War is of formal concern primarily to states parties to the multilateral instruments in question while engaged in international armed conflict among themselves, at least to the extent that the relevant instruments (or portions of them) are considered not to have entered the realm of customary international law (*i.e.*, to be unavoidably binding on all states). The treaty-imposed environmental constraints associated with non-international armed conflicts are unfortunately (though understandably) quite modest. I should note before continuing that neither the Law of War nor the associated Law of Arms Control and Disarmament—including their fundamental principles, their strengths and weaknesses, their ambiguities, and their applicability to environmental constraints—is not analyzed in the present study (for which see, *e.g.*, Falk, 2000; Goldblat, 1994; Parsons, 1998; Roberts, 2000; Schmitt, 1997–1998; Tarasofsky, 1993; Westing, 1997).

Of more immediate concern in this study than the Law of War is the self-imposed environmental constraints on a state's military sector that the state itself might adopt irrespective of its treaty commitments. To the extent that such national constraints exist, they would be found incorporated in the rules of engagement in the military manuals of a state. For example, the USA has done just that, based in part on its treaty commitments and in further part on constraints to which it is not internationally obligated (Quinn *et al.*, 2000; USDoA, 1993, chap. 5; US Navy *et al.*, 1995; for background, cf. Grunawalt, 1997). Thus, even though the USA has not as yet seen fit to become a state party to 1977 Protocol I on International Armed Conflicts (*UNTS* 17512), operational instructions by its Army legal branch nonetheless spell out the key environmental provisions of that instrument, suggesting that these largely repeat constraints to which the USA is already committed in one way or another (USDoA, 1993, chap. 5, pp 18–19). And the actual rules of engagement for its Navy and Marine Corps in fact simply incorporate some of the environmental constraints established by that instrument (US Navy *et al.*, 1995). Moreover, the US Navy and Marine Corps rules of engagement include the following important paragraph—believed to be the first in the military manual of any state that specifically requires protection of the environment during armed conflict (Roach, 1997; cf. also Quinn *et al.*, 2000)—under the heading 'Environmental Considerations' (US Navy *et al.*, 1995, ¶ 8.1.3):

> It is not unlawful to cause collateral damage to the natural environment during an attack upon a legitimate military objective. However, the commander has an affirmative obligation to avoid unnecessary damage to the environment to the extent that it is practicable to do so consistent with mission accomplishment. To that end, and as far as military requirements permit, methods or means of warfare should be employed with due regard to the protection and preservation of the natural environment. Destruction of the natural environment not necessitated by mission accomplishment and carried out wantonly is prohibited. Therefore, a commander should consider the environmental damage which will result from an attack on a legitimate objective as one of the factors during targeting analysis.

The *International Committee of the Red Cross (ICRC)* (Geneva) has long taken it upon itself to be the zealous custodian and tireless champion of the Law of War (*ICRC*, 1995). Quite recently, as part of this humanitarian task, the *ICRC* singled out environmental constraints and produced a set of model guidelines extracted primarily from all relevant portions of the Law of War (to some considerable extent, of course, from 1977 Protocol I), offering those guidelines to all states via the United Nations (as well as through its own subsequent efforts) for incorporation into their respective military manuals (*ICRC*, 1993). The *ICRC* is currently seeking to determine the extent to which states may have revised their military manuals on the basis of its environmental guidelines. At the same time it is working on a full-blown model military manual that is to incorporate those environmental guidelines.

It should be stressed at this point that national military manuals are of utmost value, even if they have not as yet incorporated environmental constraints, making their appropriate development and adoption by armed forces everywhere a matter of high priority (their global prevalence and adequacy are unfortunately not known to me). The potential efficacy of military manuals takes several forms (Reisman & Leitzau, 1991): (a) it is through a military manual that the abstractions that comprise the Law of War are translated into practical rules for application by armed forces; (b) armed forces can be exposed to a military manual in times of peace so that their contents are already ingrained in times of armed conflict; (c) a military manual converts a largely unenforceable body of international legal norms into a more readily enforceable body of national regulations; and (d) through its open publication, a military manual permits—and even invites—a military adversary to conform to reciprocal humanitarian constraints.

9.4 What Next?

In considering environmental military priorities for the future, I am less concerned with peacetime shortcomings than with wartime shortcomings; and as to the wartime shortcomings, less concerned with international armed conflicts than with non-international armed conflicts.

It is of course important that the states presently not parties to the several existing key multilateral treaties which establish wartime constraints of special environmental value be urged to rectify that dereliction. Especially important to begin with would be the following *five* instruments: (a) the *1925 Geneva Protocol* on Chemical and Biological Warfare (LNTS 2138) (as of 14 May 1998 with 60 non-parties out of 192), with the states acceding without any second-use reservation (Westing, 1985); (b) *1977 Protocol I* (as 2 March 1999 with 39 non-parties out of 192) together especially with its optional *Article 90* fact-finding commission (with 139 non-parties out of 192); (c) 1977 Protocol II on Non-international Armed Conflict (UNTS 17513) (as of 2 March 1999 with 47 non-parties out of 192); (d) the *1980 Inhumane Conventional Weapon Convention* (UNTS 22495) (as of 24 March 1999 with 119 non-parties out of 192), necessarily together with its optional Protocol II on the use of land mines (with 124 non-parties out of 192); and (e) the *1997 Anti-personnel Mine Convention* (UNTS 35597) (as of 24 March 1999 with 125 *non*-parties). Among other institutions and individuals, both socially oriented and environmentally oriented nongovernmental organizations (NGOs) might take on this lobbying task. On the other hand—with the exceptions of a proscription against nuclear weapons (Westing, 1989) and the demilitarization of protected areas (nature reserves) of outstanding universal value (Westing, 1992b)—seeking to augment the existing restraints through newly devised more stringent treaty obligations (*e.g.*, Falk, 1984; *IUCN & ICEL*, 1995; cf. also Falk, 2000), especially as regards non-international armed conflicts, should perhaps be something to consider primarily as a normative exercise and long-term objective.

Arguably of even greater immediate value than campaigning for increased treaty adoption, would be to raise the awareness in both the general public and the armed forces throughout the world of the rapid deterioration of the global biosphere. Indeed, there has in recent decades been a progressive development of environmental norms in the world community (Westing, 1996). In some instances it may thus already suffice merely to publicize the specific importance for a state to incorporate at least some environmental constraints into its rules of engagement whether or not it is a party to the relevant multilateral instruments. Nonetheless, it is difficult to over-emphasize the importance of pervasive environmental education (Westing, 1993). And as to the states parties to various components of the Law of War, it might be a useful reminder here that, among others, the four 1949 Geneva Conventions (UNTS 970–973), the two 1977 Protocols additional to them, and the 1980 Inhumane Conventional Weapon Convention all require that their content be incorporated into school curricula or otherwise disseminated.

Widespread educational efforts, both formal and informal, throughout all levels and age groups of society are especially important in the long term if the insurgent forces of the world are to adopt environmental constraints. Environmental constraints on insurgent forces will have to be self-imposed, deriving from some combination of at least *four* factors: (a) previously inculcated environmental values; (b) the need not to alienate the civilian population; (c) an attempt to minimize environmental damage to a domain over which the insurgents hope to gain control; and (d) a desire to facilitate acceptance of their legitimacy by the outside world. This is the case because on the one hand, insurgent forces are beyond the reach of domestic law; and, on the other, because existing treaty constraints dealing with non-international conflicts are purposely weak so as not to unduly undermine the national sovereignty of the states parties and also so as not to legitimize and encourage insurgencies.

An even more sweeping (and more long-term) approach to achieving suitable environmental norms than through education is suggested by a comparison of states parties with states non-parties to certain key components of the Law of War. It becomes clear from such a comparison that acceptance of the Law of War is clearly correlated with level of democratization (including human rights and governmental integrity) on the one hand and with overall stage of social and economic development on the other (Westing, 1997, pp 548–550). Successful efforts to spread democracy and support sustainable development would have the further benefit of reducing the frequency of non-international armed conflicts, which are so inherently intractable in terms of the Law of War.

To reiterate, with non-international armed conflicts having become so prevalent in recent times, the key hope for greater wartime environmental constraints will hinge not only upon the success of pervasive educational efforts at all levels, but (somewhat more indirectly) also upon the spread of democracy and the achievement of sustainable development.

9.5 In Conclusion

This contribution rests on *two* basic premises: (a) that society has by no means rejected the use of force with deadly, destructive, and disruptive intent for the ultimate resolution of conflicts, whether international or non-international; and (b) that the global biosphere is increasingly beleaguered, *inter alia*, with its natural resources and natural sinks now being utilized unsustainably. It follows from these that efforts to protect the environment cannot be restricted to the civil sector of society, but must as well embrace the military sector, both during peacetime and wartime.

Current efforts by various agencies (*e.g.*, the United Nations, NATO) to have domestic environmental protection legislation equally applicable to the civil and military sectors at least during peacetime—as is already the case in at least 19 states—should be supported.

As to wartime, it is gratifying to recognize that the great majority of states (now over three-quarters) have adopted 1949 Geneva Convention IV (UNTS 973) as well as its 1977 Additional Protocol I (UNTS 17512), which together add direct and indirect environmental constraints on the pursuit of international armed conflict. It is thus incumbent on everyone to support the efforts of the United Nations, *International Committee of the Red Cross*, and other agencies: (a) to make adoption of these two key components of International Humanitarian Law more nearly universal; (b) to have more governments incorporate the included environmental constraints into their military manuals and rules of engagement; and (c) to encourage the education mandated by those instruments.

Non-international armed conflict is as yet poorly served by International Humanitarian Law, a dilemma difficult to address owing in large part to issues of national sovereignty. Beyond working toward the more nearly universal adoption of 1977 Additional Protocol II (UNTS 17513) and the 1997 Anti-personnel Mine Convention (UNTS 35597), it is of overriding importance that widespread efforts be made to foster environmental education. It is crucial that this be done at all age levels, both in the formal and informal educational spheres, and in both the civil and military sectors. This is suggested so that the environmental norms thereby instilled will serve here regardless of the existence or acceptance of appropriate treaty obligations. Finally, democratization, the rule of law, and the achievement of sustainable development must be fostered worldwide as potent means of reducing the numbers of non-international wars.

References

ACDA. 1997. *World Military Expenditures and Arms Transfers 1996*. 25th edn. Washington: US Arms Control & Disarmament Agency, 192 pp.

Falk, R.A. 1984. Proposed convention on the crime of ecocide. In: Westing, A.H. (ed.). *Environmental Warfare: a Technical, Policy and Legal Appraisal*. London: Taylor & Francis, 107 pp: pp 45–49.

Falk, R. [A.]. 2000. The inadequacy of the existing legal approach to environmental protection in wartime. In: Austin, J.E., & Bruch, C.E. (eds). *The Environmental Consequences of War: Legal, Economic, and Scientific Perspectives*. Cambridge, UK: Cambridge University Press, 691 pp: pp 137–155 (Chap. 4).

Goldblat, J. 1994. *Arms Control: a Guide to Negotiations and Agreements*. London: Sage Publications, 772 pp.

Goodman, S.W. 1994. DoD's [US Department of Defense's] vision for environmental security. *Defense Issues* (Washington) 9(24):1–8.

Goodman, S.[W.] 1997. United States action in the field of security and the environment. In: IRIS (ed.). *Deuxièmes Conférences Stratégiques Annuelles de l'IRIS [Institut de Relations Internationales et Stratégiques]*; Paris: Documentation Française, 335 pp: pp 223–232.

Grunawalt, R.J. 1997. The JCS [Joint Chiefs of Staff] Standing Rules of Engagement: a Judge Advocate's primer. *Air Force Law Review* (Maxwell Air Force Base, AL, USA) 42:245–258.

ICRC. 1993. *Guidelines for Military Manuals and Instructions on the Protection of the Environment in Times of Armed Conflict*. Geneva: International Committee of the Red Cross, 6 pp.

Reprinted in: New York: *United Nations General Assembly* Document No. A/49/323 (19 Aug 94), 53 pp: pp 49–53.
Reprinted in: *American Journal of International Law* (Washington) 89:641–644. 1995.

ICRC. 1995. *Law of War: Prepared for Action: a Guide for Professional Soldiers*. Geneva: *International Committee of the Red Cross*, Division for Dissemination to the Armed Forces, 28 pp.

IUCN & ICEL. 1995. *Draft Convention on the Prohibition of Hostile Military Activities in Protected Areas*. Gland, Switzerland: *International Union for Conservation of Nature (IUCN) -&- International Council on Environmental Law (ICEL)*, 5 pp.

NATO. 1996. *Environmental Guidelines for the Military Sector*. Brussels: North Atlantic Treaty Organization, Committee on the Challenges of Modern Society, 54 pp.

Parsons, R.J. 1998. The fight to save the planet: U.S. armed forces, "greenkeeping," and enforcement of the law pertaining to environmental protection during armed conflict. *Georgetown International Environmental Law Review* (Washington) 10:441–500.

Quinn, J.P., Evans, R.T., & Boock, M.J. 2000. United States Navy Development of operational-environmental doctrine. In: Austin, J.E., & Bruch, C.E. (eds). *Environmental Consequences of War: Legal, Economic, and Scientific Perspectives*. Cambridge, UK: Cambridge University Press, 691 pp: pp 156–170 (Chap. 5).

Reisman, W.M., & Leitzau, W.K. 1991. Moving international law from theory to practice: the role of military manuals in effectuating the law of armed conflict. In: Robertson, H.B., Jr (ed.). *Law of Naval Operations*. Newport, RI, USA: Naval War College Press, International Law Studies, Volume 64, 540 pp: pp 1–18.

Renew America (ed.). 1995. *Today America's Forces Protect the Environment*. Washington: Renew America, 49 pp,

Roach, J.A. 1997. The laws of war and the protection of the environment. *Environment & Security* (Quebec) 1(2):53–67.

Roberts, A. 2000. The Law of War and environmental damage. In: Austin, J.E., & Bruch, C.E. (eds). *Environmental Consequences of War: Legal, Economic, and Scientific Perspectives*. Cambridge, UK: Cambridge University Press, 691 pp: pp 47–86 (Chap. 2).

Schmitt, M.N. 1997-1998. Green war: an assessment of the environmental law of international armed conflict. *Yale Journal of International Law* (New Haven, CT, USA) 22:1–109.

Sivard, R.L. 1996. *World Military and Social Expenditures 1996*. 16th edn. Washington: World Priorities, 56 pp.

Smith, D. 1997. *State of War and Peace Atlas*. 3rd edn. London: Penguin Books, 128 pp.

Tarasofsky, R.G. 1993. Legal protection of the environment during international armed conflict. *Netherlands Yearbook of International Law* (Leiden) 24:17–79.

Tillema, H.K. 1989. Foreign overt military interventions in the nuclear age. *Journal of Peace Research* (Oslo) 26:179–196,419–420.

UN. 1993. *Agenda 21: Programme of Action for Sustainable Development; Rio Declaration on Environment and Development; Statement of Forest Principles*. New York: United Nations, Publication No. DPI/1344 (Apr 93), 294 pp.

UNCSD. 1994. *Commission on Sustainable Development: Report on the Second Session (16-27 May 1994)*. New York: United Nations Economic & Social Council, Official Records, 1994, Supplement No. 13, Document No. E/1994/33/Rev.1-E/CN.17/1994/20/Rev.1, Chap.1, paragraphs 186–187.

UNEP. 1993. *Application of Environmental Norms by Military Establishments*. Nairobi: *United Nations Environment Programme*, Governing Council Decision No. 17/5 (21 May 93), 1 p.

UNEP. 1995a. *Application of Environmental Norms by Military Establishments*. Nairobi: *United Nations Environment Programme*, Governing Council Decision No. 18/29 (25 May 95), 1 p.

UNEP. 1995b. Application of Environmental Norms by Military Establishments: Report of the Executive Director. Nairobi: *United Nations Environment Programme*, Document No. UNEP/GC.18/6 + Add.1 (27 Feb & 14 May 95), 3 + 8 pp.

UNEP. 1995c. *Meeting on Military Activities and the Environment, Linköping, 27-30 June 1995: Background Paper*. Nairobi: *United Nations Environment Programme*, Document No. UNEP/MIL/2 (13 Jun 95), 17 pp.

UNEP. 1995d. *Meeting on Military Activities and the Environment, Linköping, 27-30 June 1995: Report of the Meeting*. Nairobi: *United Nations Environment Programme*, Document No. UNEP/MIL/3 (7 Jul 95), 36 pp.

UNEP. 1996a. *Sub-regional Meeting on Military Activities and the Environment, Bangkok, 26-28 June 1996: Report of the Meeting*. Nairobi: *United Nations Environment Programme*, Document No. UNEP/MIL/SEA/1 (28 Jun 96), 50 pp.

UNEP. 1996b. *Sub-regional Meeting on Military Activities and the Environment, Bangkok, 29-31 October 1996: Report of the Meeting*. Nairobi: *United Nations Environment Programme*, Document No. UNEP/MIL/SA/1 (15 Nov 96), 64 pp.

UNGA. 1995. *General and Complete Disarmament: Observance of Environmental Norms in the Drafting and Implementation of Agreements on Disarmament and Arms Control*. New York: *United Nations General Assembly*, Resolution No. 50/70 M (12 Dec 95), 1 p. [157 (85 %) in favor, 2 abstentions, 4 against, 22 absent = 185.]

UNGA. 1996. *General and Complete Disarmament: Observance of Environmental Norms in the Drafting and Implementation of Agreements on Disarmament and Arms Control*. New York: *United Nations General Assembly*, Resolution No. 51/45 E (10 Dec 96), 2 pp. [138 (75 %) in favor, 27 abstentions, 4 against, 16 absent = 185.]

UNGA. 1997. *General and Complete Disarmament: Observance of Environmental Norms in the Drafting and Implementation of Agreements on Disarmament and Arms Control*. New York: *United Nations General Assembly*, Resolution No. 52/38 E (9 Dec 97), 1 p. [160 (86 %) in favor, 6 abstentions, 0 against, 19 absent = 185.]

USDoA. 1993. *Operational Law Handbook*. Charlottesville, VA, USA: US Department of the Army, Judge Advocate General's School, Report No. JA-422(94), 281 pp.

USDoD. 1996. *Environmental Security*. Washington: US Department of Defense, Directive No. 4715.1 (24 Feb 96), 6 + 1+2 pp.

US Navy, Marine Corps, & Coast Guard. 1995. *Commander's Handbook on the Law of Naval Operations*. [3rd edn.]. Washington: US Department of the Navy, Office of the Chief of Naval Operations, Publication No. NWP 1–14 M (formerly NWP 9 (Rev. A))/FMFM 1-10/ COMDTPUB P5800.7, [113] pp.

Warrick, J. 1998. Pentagon green light: it secured exemption in warming treaty. *International Herald Tribune* (Paris) 1998(35,725):10. 10–11 Jan 98.

Westing, A.H. 1980. Warfare in a Fragile World: Military Impact on the Human Environment. London: Taylor & Francis, 249 pp.

Westing, A.H. 1982. War as a human endeavor: the high-fatality wars of the twentieth century. *Journal of Peace Research* (Oslo) 19(3):261–270.

Westing, A.H. 1985. Towards eliminating the scourge of chemical war: reflections on the occasion of the sixtieth anniversary of the Geneva Protocol. *Bulletin of Peace Proposals* [now *Security Dialogue*] (Oslo) 16(2):117–120.

Westing, A.H. 1989. Proposal for an international treaty for protection against nuclear devastation. *Bulletin of Peace Proposals* [now *Security Dialogue*] (Oslo) 20(4):435–436.

Westing, A.H. 1992a. Environmental dimensions of maritime security. In: Goldblat, J. (ed.). *Maritime Security: the Building of Confidence*. Geneva: United Nations Institute for Disarmament Research, Document No. UNIDIR/92/89, 159 pp: pp 91–102.

Westing, A.H. 1992b. Protected natural areas and the military. *Environmental Conservation* (Cambridge, UK) 19(4):343–348.

Westing, A.H. 1993. Global need for environmental education. *Environment*, (Washington) 35(7):4–5,45.

Westing, A.H. 1996. Core values for sustainable development. *Environmental Conservation* (Cambridge, UK) 23(3):218–225.

Westing, A.H. 1997. Environmental protection from wartime damage: the role of international law. In: Gleditsch, N.P. (ed.). *Conflict and the Environment*. Dordrecht, Netherlands: Kluwer Academic Publishers, 598 pp: pp 535–553.

Glossary

# = Number	[The thus numbered references throughout the text are provided in Chap. 3.]
AAAS	American Association for the Advancement of Science, Washington, DC, USA (1848–)
ICRC	International Committee of the Red Cross, Geneva, Switzerland (1863–)
Agent Blue	A water-soluble ca 3:1 mixture of sodium dimethyl arsenate (sodium cacodylate) and dimethyl arsinic (cacodylic) acid
Agent Orange	An oil-soluble ca 1:1 mixture of 2,4-dichlorophenoxy acetic acid (2,4-D) and 2,4,5-trichlorophenoxy acetic acid (2,4,5-T), the latter containing minute amounts of dioxin
Agent White	A water-soluble ca 4:1 mixture of 2,4-dichlorophenoxy acetic acid (2,4-D) and 4-amino-3,5,6-trichloropicolinic acid (picloram)
IUCN	International Union for Conservation of Nature, Gland, Switzerland (1948–)
LNTS	*League of Nations Treaty Series*, New York, NY, USA (1920–1946) [cf. 'UNTS' below]
NATO	North Atlantic Treaty Organization, Brussels, Belgium (1949–)
PRIO	Peace Research Institute Oslo, Oslo, Norway (1959–)
SIPRI	Stockholm International Peace Research Institute, Stockholm, Sweden (1966–)
2,4-D	2,4-dichlorophenoxy acetic acid [cf. 'Agent Orange' and 'Agent White' above]
2,4,5-T	2,4,5-trichlorophenoxy acetic acid [cf. 'Agent Orange' above]
UV-B	Ultraviolet-B, that portion of the non-ionizing electromagnetic radiation spectrum in the wavelength range of ca 280–315 nanometers (nm)

A. H. Westing, *Arthur H. Westing*, SpringerBriefs on Pioneers in Science and Practice 1, DOI: 10.1007/978-3-642-31322-6,

UNCSD	United Nations Commission on Sustainable Development, New York, NY, USA (1992–)
UNEP	United Nations Environment Programme, Nairobi, Kenya (1972–)
UNESCO	United Nations Educational, Scientific and Cultural Organization, Paris, France (1948–)
UNIDIR	United Nations Institute for Disarmament Research, Geneva, Switzerland (1980–)
UNTS	*United Nations Treaty Series*, New York, NY, USA (1946–) [cf. 'LNTS' above]

Units of Measure

Standard International (SI) prefixes used in the text

T (Tera) = 10^{12}	c (centi) = 10^{-2}
G (Giga) = 10^{9}	m (milli) = 10^{-3}
M (Mega) = 10^{6}	μ (micro) = 10^{-6}
k (kilo) = 10^{3}	n (nano) = 10^{-9}
h (hecto) = 10^{2}	

are (a)	100 square meters = 1076 square feet [cf. 'hectare' below]
cubic meter (m^3)	[cf. 'meter, cubic' below]
degree Celsius (°C)	a measure of temperature. To convert degrees Celsius to degrees Fahrenheit (°F), first multiply by 1.8, then add 32
meter, cubic (m^3)	10^3 liters (L) = 264.2 US gallons = 220.0 British gallons = 6.290 US oil barrels = 0.000811 acre-foot
gram (g)	0.03527 ounce = 0.002205 pound [cf. 'kilogram' below]
hectare (ha)	10^4 square meters = 0.01 square kilometer = 2.471 acres
joule (J)	0.2388 calorie = 0.0009486 British thermal unit (BTU)
kilogram (kg)	2.205 pounds
kilogram per hectare (kg/ha)	0.892 pound per acre
kilometer (km)	10^3 meters = 0.621 statute mile = 0.540 nautical mile
kilometer, square (km^2)	10^6 square meters = 100 hectares = 247.1 acres = 0.386 square statute mile

A. H. Westing, *Arthur H. Westing*, SpringerBriefs on Pioneers in Science and Practice 1, DOI: 10.1007/978-3-642-31322-6,

kilopascal (kPa)	0.00987 atmosphere = 0.145 pound per square inch
liter (L) [formerly 'l']	0.001 cubic meter (10^{-3} m^3) = 0.2642 US gallon = 0.2200 British gallon
meter (m)	3.281 feet
meter, cubic (m^3)	10^3 liters = 264.2 US gallons = 220.0 British gallons = 6.290 US oil barrels = 0.000811 acre-foot
pascal (Pa)	[cf. 'kilopascal' above]
Röntgen (R)	a measure of radiation exposure, with 1 R being approximately equivalent to an absorbed dose of 1 rad or 10^{-2} gray (Gy)
tonne (t)	10^3 kilograms = 1.102 US (short) tons (T) = 0.984 British (long) ton
watt (W)	1 joule per second (J/s). The designation 'e', as in 'W(e)', indicates that the energy is in the form of electricity.